W9-DFR-249

FROM ERROR-CORRECTING CODES THROUGH SPHERE PACKINGS TO SIMPLE GROUPS

By

THOMAS M. THOMPSON

THE

CARUS MATHEMATICAL MONOGRAPHS

Published by

THE MATHEMATICAL ASSOCIATION OF AMERICA

THE CARUS MATHEMATICAL MONOGRAPHS are an expression of the desire of Mrs. Mary Hegeler Carus, and of her son, Dr. Edward H. Carus, to contribute to the dissemination of mathematical knowledge by making accessible at nominal cost a series of expository presentations of the best thoughts and keenest researches in pure and applied mathematics. The publication of the first four of these monographs was made possible by a notable gift to the Mathematical Association of America by Mrs. Carus as sole trustee of the Edward C. Hegeler Trust Fund. The sales from these have resulted in the Carus Monograph Fund, and the Mathematical Association has used this as a revolving book fund to publish the succeeding monographs.

The expositions of mathematical subjects which the monographs contain are set forth in a manner comprehensible not only to teachers and students specializing in mathematics, but also to scientific workers in other fields, and especially to the wide circle of thoughtful people who, having a moderate acquaintance with elementary mathematics, wish to extend their knowledge without prolonged and critical study of the mathematical journals and treatises. The scope of this series includes also historical and biographical monographs.

The following monographs have been published:

No. 1. Calculus of Variations, by G. A. BLISS

No. 2. Analytic Functions of a Complex Variable, by D. R. CURTISS

No. 3. Mathematical Statistics, by H. L. RIETZ

No. 4. Projective Geometry, by J. W. YOUNG

No. 5. A History of Mathematics in America before 1900, by D. E. SMITH and JEKUTHIEL GINSBURG (out of print)

No. 6. Fourier Series and Orthogonal Polynomials, by DUNHAM JACKSON

No. 7. Vectors and Matrices, by C. C. MACDUFFEE

No. 8. Rings and Ideals, by N. H. MCCOY

No. 9. The Theory of Algebraic Numbers, Second edition, by HARRY POLLARD and HAROLD G. DIAMOND

No. 10. The Arithmetic Theory of Quadratic Forms, by B. W. JONES

No. 11. Irrational Numbers, by IVAN NIVEN

No. 12. Statistical Independence in Probability, Analysis and Number Theory, by MARK KAC

No. 13. A Primer of Real Functions, Third edition, by RALPH P. BOAS, JR.

No. 14. Combinatorial Mathematics, by HERBERT JOHN RYSER

No. 15. Noncommutative Rings, by I. N. HERSTEIN

No. 16. Dedekind Sums, by HANS RADEMACHER and EMIL GROSSWALD

No. 17. The Schwarz Function and its Applications, by PHILIP J. DAVIS

No. 18. Celestial Mechanics, by HARRY POLLARD

No. 19. Field Theory and Its Classical Problems, by CHARLES ROBERT HADLOCK

No. 20. The Generalized Riemann Integral, by ROBERT M. McLEOD

No. 21. From Error-Correcting Codes through Sphere Packings to Simple Groups, by THOMAS M. THOMPSON.

The Carus Mathematical Monographs

NUMBER TWENTY-ONE

FROM ERROR-CORRECTING CODES THROUGH SPHERE PACKINGS TO SIMPLE GROUPS

By

THOMAS M. THOMPSON
Walla Walla College

Published and Distributed by
THE MATHEMATICAL ASSOCIATION OF AMERICA

Library of Congress Catalog Card Number 82-062784

Complete Set ISBN 0-88385-000-1
Vol. 21 ISBN 0-88385-023-0

Printed in the United States of America

Current printing (last digit):

10 9 8 7 6 5 4 3 2 1

PREFACE

Back in 1947 Richard W. Hamming had access to a computer only on weekends. Some three decades later he recalled his frustration over its perverse behavior:

> Two weekends in a row I came in and found that all my stuff had been dumped and nothing was done. . . . And so I said, 'Damn it, if the machine can detect an error, why can't it locate the position of the error and correct it?' [**56**, Tape 2]

That question initiated the development of error-correcting codes. The offending computer, a mechanical relay model at Bell Telephone Laboratories, always came to an angry stop and switched to the next program whenever it detected an error. This behavior impelled Hamming, a pure mathematician with an applied bent, to devise the first error-correcting code.

We shall follow the devious trail that wends its way through a quarter-century of mathematics, starting with Hamming's work, which led almost immediately to that of M. J. E. Golay. The latter sparked, some twelve years later, giant steps in the packing of congruent spheres by John Leech, which, in turn, branched off through the work of J. H. Conway, into the field of simple groups. By tracing some of the twists, turns, switchbacks and dead ends of this path, we hope to provide a small window on the history of mathematics of the twentieth century. How historians ultimately will treat this golden age of mathematical creativity we cannot even guess. But one claim we can make on the basis of this study is certain: The record will not be neat and tidy. It will not divide naturally and easily into chapters with such convenient labels as "Combinatorics" or "Geometry" or "Algebra." The more the history of these times is artificially divided into such categories, the more it will mislead the inquirers who ask, "How did it all happen?"

This century presents a paradox to the historian. The high affairs of state, the spectacular in sports, and the splash of celebrities are recorded on film and videotape, in books, newspapers and magazines, in detail unparalleled in the past. However, the electronic revolution that has helped make such thorough documentation possible, renders the doings of the less conspicuous more ephemeral. A telephone conversation leaves no trace, while letters used to be saved and published in many a correspondence or Briefwechsel. Who, anymore, spends time on a diary? Even the ease of jet travel helps undermine the written record. Granted, journals do publish the research, but high page costs encourage a terse expository style. There is little room to recount the origin of the paper or to reveal what the "theorem is all about." Survey articles, while attempting to compensate for this omission, can hardly provide this background, since their starting point is usually the published literature. As George Pólya said in a recently published interview [**2**, pp. 16, 17]:

> ...Among the old mathematicians, I was most influenced by Euler and mostly because Euler did something that no other great mathematician of his stature did. He explained how he found his results and I was deeply interested in that. It has to do with my interest in problem solving. ...And as I came to mathematics and learned something of it, I thought: Well, it is so, I see, the proof seems to be conclusive, but how can people find such results? My difficulty in mathematics: How was it discovered?

Mathematicians are not given to writing and certainly seldom to autobiographical writing. Later centuries may know as little of the leading mathematicians of this age as we know of Shakespeare. They may be able to uncover as little of how the triumphs of this age were achieved as we can of the manner of constructing the pyramids of ancient Egypt. The trail we trace in this book would have disappeared in another thirty or forty years. Fortunately, the mathematicians who followed this route through their contributions are still very much alive. One gladly participated in many hours of taped interviews. Another, who abhorred the telephone, was generous with his written replies to my letters. A third, who is notorious for never replying to a letter, was quite willing to talk at length on the transatlantic phone. Indeed, their enthusiasm and their belief in the merit of this project helped sustain my efforts over the two years it took to clarify the history and the mathematics.

But even with all this encouragement and help, I was not able to paint a complete picture. There is still disagreement on some of the historical facts. On others, memories could not fill all the gaps. In even one case, antipathy for the entire project, indeed on the history of mathematics in any form, closed the door on the corroboration of certain details.

I was myself many times astonished by the quirks and flukes, the coincidences that lay behind the discoveries. One mathematician was goaded by a contrary machine. A physicist was inspired by a rather obscure example tucked away in the middle of a paper. Another mathematician believed on the basis of quite flimsy evidence that a certain group of symmetries would contain a large simple group. And yet another, who studied this group and determined its order, was so certain that it was simple he didn't even bother to stop and verify that it was.

The trail of mathematical discovery is strewn with gambles, hunches, oddities of character, luck, idiosyncrasies of background, and fortunate encounters. Chaos and spontaneity, not austere order, mark the way much of mathematics is done. The elegance of the printed article conceals the circumstances that precipitated the results. The history of mathematics is a story of people and their guesses, misfortune and struggle, not merely a list of theorems and their proofs.

In this book, we have taken a somewhat novel approach. We describe the mathematics in complete detail and its origin and evolution as well. Where the journal article is concise, because it was written with a narrow audience in mind, we have supplied the missing steps. Thus, any mathematician with even a casual acquaintance with vector spaces and groups should be able to follow each mathematical step. In fact, I wrote each page so that an upper-division student could understand it.

Along the way we will see the interplay between applied and pure mathematics, the importance of direct personal contact, and of conferences and travel. Implicitly, we can deduce the danger of isolation, of being outside the opportunity of chance contact, of missing the stimulation that comes from browsing through new books. And we will see how mathematics is discovered and lost, only to be rediscovered.

While the reader may draw many a moral from our tale, I hope that the story is of interest for its own sake. Moreover, I hope that it may inspire others, participants or observers, to preserve the true and complete record of our mathematical times. I have followed only one thread in the intricate tapestry. Many others must be carefully scrutinized before we can weave the complex design of the whole. The eventual record will serve not only the purposes of history, but also mathematics itself, providing students now and in future generations a deeper insight into the mathematics that is their heritage.

The book consists of three chapters, each describing one topic: error-correcting codes, sphere packings and simple groups. Their connection was the basis of our interest.

Chapter 1 begins with an introduction to coding and leads directly into the early, unpublished work of Hamming. This, in turn, is followed by the introduction of Golay, whose work was inspired by that of Hamming through an example cited by C. E. Shannon in his classical treatise on information theory. The priority controversy between Hamming and Golay which ensued because of Hamming's delay in publishing due to patent considerations (Bell Laboratories was actually able to patent his mathematical code!) is discussed in the last section.

Chapter 2 describes the bridge between the Golay codes and a surprisingly dense sphere packing in twenty-four dimensional Euclidean space E^{24} discovered by John Leech. This packing actually evolved in two steps, the second being the trigger for the discovery of new large simple groups, the topic of Chapter 3. The history of all the influences that led to Leech's discovery remains obscured. Certainly, Golay's work played a major role. But Leech also referenced L. J. Paige and E. H. Spanier whose work, while reminiscent of that of Golay, actually dealt with some well-known simple groups, those of E. Mathieu which had been discovered nearly one hundred years earlier.

Leech's lattice in E^{24} is the focal point of Chapter 3. Leech wanted to know its symmetries. Unable to answer this question and convinced that the answer would "repay investigation" [**81**], Leech tried to interest others in his question. Finally, J. H. Conway, informed about the lattice by John McKay and challenged to work on it by John Thompson, answered the question. The result was the addition of three new simple groups to the growing list of new sporadic simple groups.

Two of the appendices may be of special interest to the reader. Appendix 1 summarizes the densest known sphere packings, and Appendix 6 contains the now complete list of sporadic simple groups.

Much credit for the oral history material in this book goes to those who actually made the history. Without their cooperation, the task could not have been completed. Professor John Leech, besides carefully reviewing the chapter on his work, offered valuable comments on the other two chapters. The University of California at Davis and Walla Walla College in College Place, Washington, provided the necessary financial assistance. I am deeply indebted to Professors C. R. Borges, G. D. Chakerian and especially to S. K. Stein for their inspiration, encouragement, suggestions and careful reading of the manuscript, and to Bonnie-Jean McNiel who braved a mass of ever-changing notation to type it.

THOMAS M. THOMPSON

College Place, Washington
February, 1983

CONTENTS

CHAPTER 1

THE ORIGIN OF ERROR-CORRECTING CODES

Section 1. An Introduction to Coding

Richard W. Hamming's encounter with the Bell Telephone Laboratories' mechanical relay computer in 1947 (quoted in the Preface) initiated what has come to be known as coding theory. In this chapter we will trace these origins of coding theory and develop enough of the theory itself to prepare for Leech's work in sphere packing which appears in Chapter 2.

Communication is imperfect. Even if a message is accurately stated, it may be garbled during transmission; and the consequences of a mistake in the interpretation of a financial, diplomatic, military or other message may be unfortunate. (Hamming's work was goaded by mistakes which occurred internally in a computer.) As a result, when a message travels from an information source to a destination, both the sender and the receiver would like assurance that the received message either is free of errors or, if it contains errors, those errors will be detectable. Ideally, in case an error is detected, the receiver would like to be able to correct it and recover the original message.

Our interest will lie mainly in the transmission of strings of 0's and 1's of length n, which are called (binary) *n-blocks.* This may, at first glance, seem an artificial restriction. But in fact the entire twenty-six letters of the alphabet can be coded using 5-blocks since there are thirty-two arrangements of five 0's and 1's. More generally, there are 2^n n-blocks available to form messages.

A sequence of such n-blocks will constitute a *message*. For example, one message of six 3-blocks is

$$001 \quad 101 \quad 110 \quad 110 \quad 010 \quad 111.$$

When errors never occur in transmission, all such n-blocks may be used to form messages without risk of misinterpretation. However, in case errors can occur, the block received may differ from the one sent. One way to compensate for this is to send only blocks that differ from each other so much that one or two errors cannot change one of the selected blocks into another. This restriction requires the use of longer blocks in order to have the same variety of messages possible with the 2^n n-blocks. Of course, this will reduce the amount of information sent per unit time: The gain in reliability is paid for by accepting a slower rate of transmission. A set of selected n-blocks, called *codewords*, will make up a (binary) *code*.

The simplest code has just the two codewords, 0 and 1. In this case no errors in transmission can be detected, since any error changes one codeword into the other. The code $\{00, 11\}$, formed by simply repeating the digits of the old codewords, will detect single errors since any single error in a codeword changes it to a noncodeword. For example, 00 could be sent and 01 received. However, two errors in the transmission of a single codeword will go undetected since the received word is also a codeword.

By repeating each of the binary symbols $n-1$ times, we form the *repetition code* consisting of two words of length n, the all-zero word and the all-one word. Up to $n-1$ errors in transmission are detectable. The first digit of each codeword is the *message digit* while the following $n-1$ digits are the *check digits*.

The repetition codes allow the sending of only two different codewords. This restriction can be avoided by using a *block repetition code* whose codewords consist of s copies of all possible r-blocks of binary digits instead of n copies of a single binary digit. There are 2^r words of length $n = rs$ in this code. Take, for example, the case $r = 4$ and $s = 2$. Form the code of sixteen words of length eight where the first four digits coincide with the last four digits. As in the case of the repetition code of length two, all single errors are detectable. For instance, an error in the eighth digit of the codeword 11001100 yields the 8-block 11001101, which is not a codeword. We represent this transmission pictorially as:

$$11001100 \longrightarrow 11001101$$

(The horizontal arrow is read "is received as.") However, two errors are not always detectable as the following shows:

$$11001100 \longrightarrow 11101110$$

This simple generalization of the repetition codes is equivalent to repeating a given message of r-blocks s times.

A binary block code of length n having r message digits is called an *(n, r) binary block code.* We will refer to it as an (n, r) code. Usually the r message digits occupy the first r positions of a codeword.

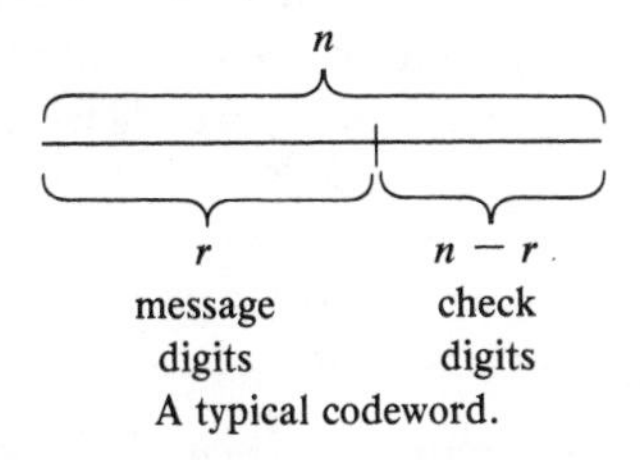

A typical codeword.

For example, the repetition code would be an $(n, 1)$ code while the block repetition code would be an (rs, r) code.

Suppose that the probability that a certain message will be

sent is p and that no errors in transmission can occur. If p is close to 1, then no one is "surprised" very much by receiving the message and thus it conveys, in some sense, little information. On the other hand, if p is very small, the "surprise" is much larger and in the same sense, much more information is received. The *information* contained in such a message is given by

$$\log_b\left(\frac{1}{p}\right)$$

for some appropriate base b. This not only agrees with the notion of "surprise," but it is additive in the following sense: If two independent messages occur with positive probabilities p_1 and p_2, then the probability that both occur is the product p_1p_2. In that case, the information sent is

$$\log_b\left(\frac{1}{p_1p_2}\right) = \log_b\left(\frac{1}{p_1}\right) + \log_b\left(\frac{1}{p_2}\right),$$

i.e., the sum of the information of the individual messages. If base 2 logarithms are used, the unit of information is the *bit*. This means that equally likely binary digits carry $\log_2\left(\frac{1}{1/2}\right) =$ 1 bit of information each. The information carried by one codeword of an (n, r) code with w (equally likely) codewords is thus $\log_2\left(\frac{1}{1/w}\right) = \log_2 w$ bits. In case $w = 2^r$, this reduces to r, which is just the number of message digits.

The *information rate* of an (n, r) code with w codewords is the quotient

$$\frac{\log_2 w}{n}.$$

In case $w = 2^r$, this reduces to r/n. This rate records how much information (in bits) is carried on the average per symbol sent. For instance, the information rate of the $(n, 1)$ repetition code is $1/n$, while that of the (rs, r) block repetition code is r/rs, which is $1/s$. In each case the addition of check digits automatically lowered the information rate.

The goal of coding theory is to devise ways to send messages quickly and accurately. Achieving accuracy over a noisier channel requires more error detection which means more check digits per codeword and thus longer codewords. This slows the flow of information. The cumbersome repetition codes turn out to be very slow but fairly reliable. The concept of "parity" will help realize both goals, speed and reliability, at the same time.

An n-block of binary digits has *even parity* if the sum modulo 2 of the digits is zero (in other words, there are an even number of 1's). Otherwise the n-block has *odd parity.* The summing of the digits is the *parity check.*

With the aid of the parity check, we can devise an efficient single-error-detecting code. For example, to each of the sixteen possible 4-blocks, annex a fifth digit so that the 5-blocks formed have even parity. Thus

$$1100$$

is encoded as

$$11000,$$

while

$$1101$$

becomes

$$11011.$$

Any single error in the transmission of a 5-block makes the parity of that codeword odd and we say that the parity check *fails.* For example:

$$11000 \longrightarrow 11010$$

indicates an error since $1 + 1 + 0 + 1 + 0 \equiv 1 \pmod 2$. Any two errors escape detection since the parity of the received word is still even. Like the (8, 4) block repetition code, this (5, 4) single-parity-check code detects single errors. However, its information rate is greater, being $4/5 = 0.80$ as compared with $4/8 = 0.50$ for the (8, 4) code. Since single-error detecting codes with four message digits per codeword must have at least one check digit, the (5, 4) code has the highest rate possible among single-error-detecting codes with four message digits.

Finding high rate codes is not the only problem. One has to consider the channel over which the messages pass. For example, the *binary symmetric channel* is illustrated in Figure 1.1.

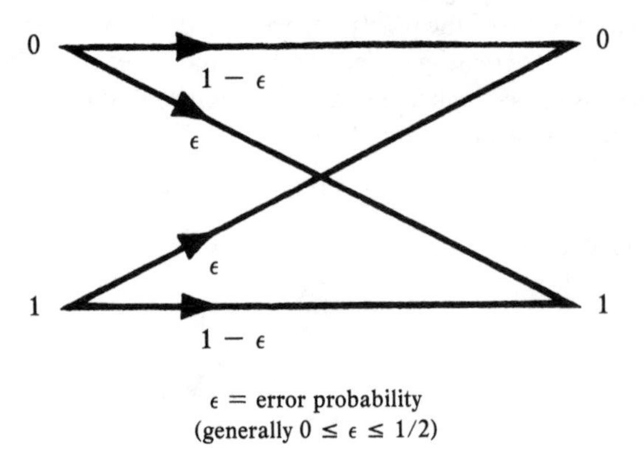

ϵ = error probability
(generally $0 \leq \epsilon \leq 1/2$)

FIG. 1.1. The binary symmetric channel.

How much information can pass over this channel? If $\epsilon = 0$, then for one bit of information sent, one could expect to receive one bit. Suppose that $\epsilon > 0$ and that one bit of information is being sent. If an error occurs, then the amount of information lost in that error is $\log_2\left(\frac{1}{\epsilon}\right)$. On the average this is $\epsilon \log_2 \frac{1}{\epsilon}$. Because of the ambiguity ($\epsilon > 0$), even if an error does not occur, $(1 - \epsilon) \log_2\left(\frac{1}{1-\epsilon}\right)$ bits on the average are lost. The total amount lost is the sum of the two. This means that the amount of information the binary symmetric channel is able to pass or its *channel capacity* (in bits per digit) is given by:

$$\begin{aligned} C(\epsilon) &= 1 - \left[\epsilon \log_2\left(\frac{1}{\epsilon}\right) + (1-\epsilon)\log_2\left(\frac{1}{1-\epsilon}\right)\right] \\ &= 1 + \epsilon \log_2 \epsilon + (1-\epsilon)\log_2(1-\epsilon). \end{aligned}$$

The capacity is a measure, between 0 and 1, of the effect of noise on the channel. It is easy to check that $C(0) = \lim_{\epsilon \to 0} C(\epsilon) = 1$ while $C(1/2) = 0$. The latter simply means that if there is a fifty-fifty chance of error for each binary digit sent, no information can be sent over the channel. Claude E. Shannon showed in 1948 [**108,** p. 411 ff.] that information flows at nearly the rate $C(\epsilon)$ with probability of error arbitrarily small. For binary codes this means that, given arbitrary $\delta > 0$ and $R < C(\epsilon)$, there exist (n, r) codes, with n sufficiently large so that $r/n \geq R$ and the probability of incorrectly decoding a word at the receiving end is less than δ. Unfortunately, no constructive proof exists. It is also known that if $C(\epsilon)$ is exceeded by the code rate, the probability of

TABLE 1.2

Decimal Digit	Codeword
1	11000
2	10100
3	01100
4	10010
5	01010
6	00110
7	10001
8	01001
9	00101
0	00011

error can no longer be made arbitrarily small. For further information, see [**1**], [**59**] or [**85**].

In the early 1940's, Bell Telephone Laboratories used a code similar to the (5, 4) single-parity-check code in two of its relay computers [**6,** p. 349]. The codewords were the ten possible binary 5-blocks with exactly two 1's. These represented the ten decimal digits. Table 1.2 displays the correspondence for this so-called *two-out-of-five code* [**58**]. Each decimal digit was stored on a group of five relays by having exactly two of them switched on in the proper locations [**3,** p. 3]. Any single relay malfunction was immediately detectable, since in the case of such a malfunction exactly two relays would no longer operate, but, instead, either one or three.

A later model relay computer used a *three-out-of-five code* [**3,** p. 3]. This code also has ten codewords since $\binom{5}{3} = 10$. Mathematically, it is obtained from the two-out-of-five code by interchanging the 0's and 1's; physically, it was quite different, since three relays had to be activated instead of two. In case of parity check failure, the checking circuits automatically and immediately halted the machine.

We now turn from error detection to error correction. It is one thing to know that a particular message is wrong, but it is quite another to be able to reconstruct the correct message from the wrong message. To study this problem, we shall take the geometric point of view introduced by Hamming in his 1950 paper on error-correcting codes [**54,** pp. 154–156]. He first considered the unit cube in Euclidean n-dimensional space, E^n, whose vertices are the 2^n n-tuples of 0's and 1's. A binary code with words of length n is then a certain subset of the vertices of this cube. The darkened circles in Figure 1.3 illustrate a code with four codewords of length three. The four darkened vertices are the codewords 000, 011, 101 and 110 formed by the addition of a parity check digit to each pair of binary digits. This (3, 2) code detects single errors. Figure 1.4 shows the three possible 3-blocks formed by a single error

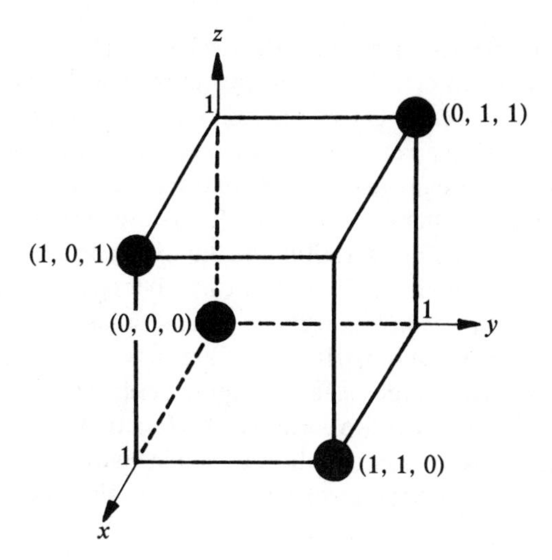

Fig. 1.3

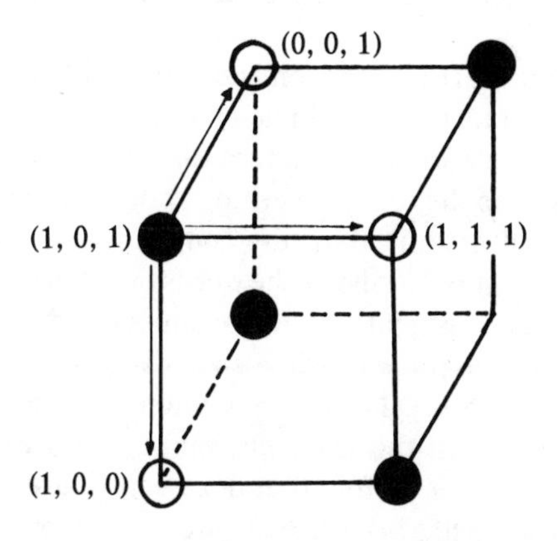

Fig. 1.4

in the transmission of the codeword 101. This figure also shows why two errors in a codeword are not detectable: a second error sends the result of the first error to another codeword. One error in a codeword changes one coordinate and two errors change two coordinates; for any positive integer e, e errors alter e coordinates. This observation of Hamming motivated his definition of the *distance*, D, between two n-blocks of binary digits as the number of coordinates in which the corresponding vertices differ. For instance, $D(101, 011) = 2$, while $D(101, 010) = 3$. In terms of the geometry of the cube, this distance is just the number of edges in a shortest path between the two vertices. This function D, called the *Hamming distance*, a standard notion in coding theory, satisfies the usual definition of a metric. For n-blocks x, y and z,

$$D(x, y) \geq 0 \text{ and } D(x, y) = 0 \Leftrightarrow x = y$$
$$D(x, y) = D(y, x)$$
and
$$D(x, z) \leq D(x, y) + D(y, z).$$

The *minimum distance* of a code is the minimum of all the distances between two nonidentical codewords of the code. For the (3, 2) single-parity-check code shown in Figure 1.3, it is two.

While single-error detection is possible with a code whose minimum distance is two, single-error correction is only possible with a code whose minimum distance is at least three. One example is the (3, 1) repetition code pictured in Figure 1.5. (We will assume that any code contains the all-zero codeword.) Figure 1.6 shows the words that can be received if exactly one error is made in the transmission of the codeword 000. Each of the three possible errors still is a distance two from the codeword 111. If a maximum of one error is assumed to occur in the transmission of a single codeword, then the error can be found and we can figure out exactly which codeword has been sent. However, the correcting ability disappears if two errors occur in the transmission of one

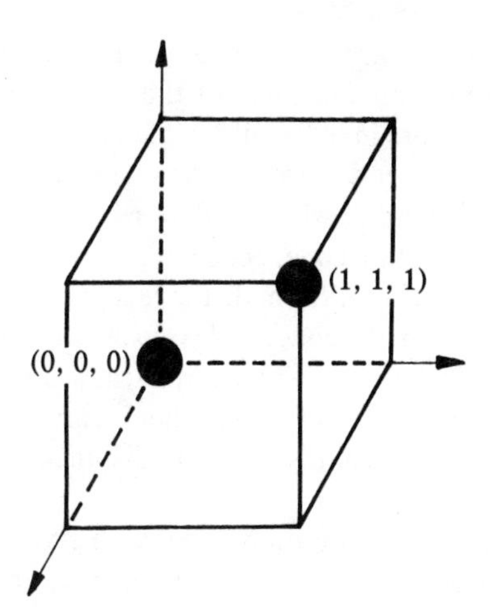

FIG. 1.5

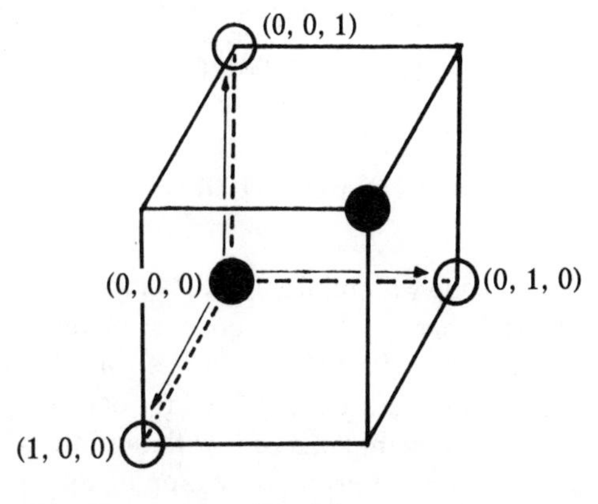

FIG. 1.6

codeword. In this case a received word resulting from two errors is within a distance one of the other codeword. For example, two errors in 000 could produce 110, which can be obtained from the codeword 111 by one error. The receiver has no way of deciding which codeword was sent.

The (4, 1) repetition code, consisting of the two codewords 0000 and 1111, has minimum distance four. Any single error in the transmission of a codeword is easily corrected since the received word is at distance three from the second codeword. Two errors can be detected but they cannot be corrected since the received word this time is at distance two from both codewords.

The (16, 4) block repetition code, with sixteen instead of two codewords, has the same minimum distance (four) as is illustrated by the codewords

1100110011001100

and

1110111011101110.

However, the distance between some of its sixteen codewords is as great as sixteen; for example, the distance between

1100110011001100

and

0011001100110011

is sixteen.

In general, a code with minimum distance d will detect up to $[d/2]$ errors (where [ ] is the greatest integer function) and simultaneously correct $[(d-1)/2]$ errors. Some of the correction capability may be sacrificed for additional detection capability. For example, a code with minimum distance d can detect up to $d-1$ errors by scrapping all correction.

Hamming's geometrical model further suggests the notion

of a perfect code. For nonnegative integer ϵ, Hamming defined a *sphere of radius* ϵ, centered on a vertex of the unit cube in E^n, as the set of all vertices of the cube at a Hamming distance ϵ from the given vertex [**54,** p. 155]. The term, as it is used now, includes all vertices within ϵ units of the given vertex so that each sphere contains $\binom{n}{0} + \binom{n}{1} + \cdots + \binom{n}{\epsilon}$ vertices of the cube. Figure 1.7 shows the 1-spheres centered at the two codewords of the (3, 1) repetition code. Note that the two 1-spheres exhaust the set of all vertices of the cube and do not overlap.

A code of length n is called *perfect* if there is a nonnegative integer ϵ such that:

1. The ϵ-spheres centered on the codeword vertices are pairwise disjoint.
2. Each vertex of the n-cube is contained in some ϵ-sphere.

Such codes are also called *close-packed* or *lossless.* The repetition codes with n odd are perfect. For such codes, take $\epsilon = (n - 1)/2$. This is shown for the (3, 1) code in Figure 1.7. In such codes, each of the two ϵ-spheres contains 2^{n-1} vertices; the one centered on the all-zero word contains all the vertices with less than $n/2$ 1's while the one centered on the all-one codeword contains all those with more than $n/2$ 1's. The (2, 1) repetition code, pictured in Figure 1.8, is not perfect since 0-spheres centered at codewords miss two vertices while 1-spheres overlap. What happens with $n = 4$? The 1-spheres don't cover while the 2-spheres overlap.

Section 2. The Work of Hamming

We have already shown that the (8, 4) block repetition code, with rate 0.50, does not have the highest information rate for a single-error-detecting code with each codeword having four message digits. The (5, 4) single-parity-check code, with rate

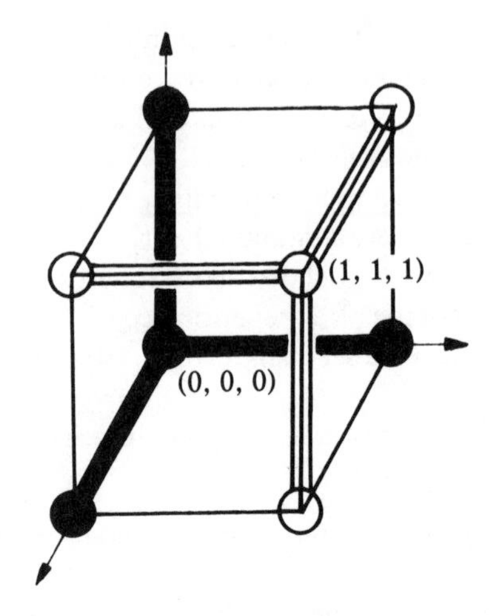

FIG. 1.7

0.80, is much better. Can a similar device be used to improve single-error-correcting codes? For example, the (12, 4) block repetition code has minimum distance three and thus is single-error-correcting. But its rate is only $4/12 = 0.33$.* Can single-error-correction for a code where each codeword has four message digits be accomplished by adding fewer than eight check digits per codeword?

This question was answered indirectly in July, 1948, by Shannon in his fundamental paper on information theory, "A Mathematical Theory of Communication" [**108**]. Forty pages into this paper appears an extremely efficient (7, 4) single-error-correcting code, described in a mere eleven lines, and attributed to Hamming. Its information rate is $4/7 = 0.57$. Hamming, because of patent considerations, was not able to publish the details or generalizations of this code

*All decimals will appear in truncated, not rounded, form.

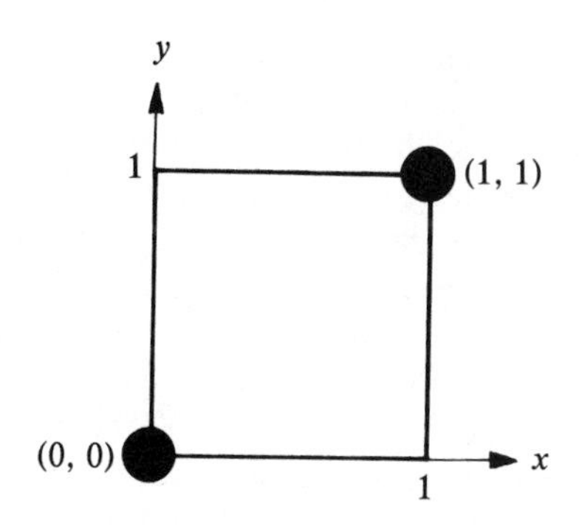

Fig. 1.8

until nearly two years later [**57**]. Shannon's early announcement not only marked the beginning of a new field of mathematics—error-correcting codes—but also precipitated a priority dispute between Hamming and Marcel J. E. Golay who first published on coding in 1949 [**13,** pp. 2, 3], [**40**], [**43**]. It should be noted that although Shannon's paper was the first to include any reference to error-correcting codes in coding terminology, some results in coding were actually discovered earlier in quite different contexts: for example, in 1942 the work of R. A. Fisher [**36**] in the design of experiments. This and other precursors have been noted in [**13,** p. 8], [**17,** pp. 7, 8].

Hamming did not develop his codes simply out of curiosity. When he left the atomic bomb project at Los Alamos in July, 1946, to go to Bell Laboratories, he carried with him an interest in large-scale computing machines [**11**], [**56,** tape 2]. Although his undergraduate and graduate training was in pure mathematics, he had an applied bent, which led him to write his doctoral thesis on differential equations.

Bell Laboratories had been developing relay computing machines since George R. Stibitz designed and built his "Complex Number Computer" at the Laboratories in 1937, a machine later named the Model I [**6,** pp. 344, 348]. Before Hamming's arrival at the Laboratories, one Model V machine had already been built for the National Advisory Committee on Aeronautics and the second was nearly complete. The Ballistics Research Laboratories in Aberdeen, Maryland, was to receive the latter machine, which Hamming was eventually to use in his research [**6,** p. 350], [**58**].

The Model V, the penultimate relay computer to be built by Bell, was the largest in the series. Containing over nine thousand relays and over fifty pieces of teletype apparatus, it occupied about one thousand square feet of floor space and weighed some ten tons [**3,** p. 1]. Incidentally, the same computing power three decades later is to be found in the more sophisticated hand-held calculators. The Model V consisted of two computing sections which could be joined for the treatment of large problems. More sections could have been added if needed [**4,** p. 81], [**125,** Figure 12]. By today's standards the machine was very slow. For example, it took about three-and-one-half hours to solve a system of thirteen linear equations by the elimination process [**4,** p. 81]. Now such a computation can be executed in less than a minute on a micro-computer, which occupies less than a cubic foot of space.

Besides the three-out-of-five codes already mentioned, the Model V also used various k-out-of-six codes, particularly for input and output [**125,** p. 49]. These were transmitted essentially in blocks and permitted extensive self-checking. For example, the input was entered in the machine via a punched paper tape, seven-eighths of an inch wide, which had up to six holes per row [**3,** p. 8]. Each row was read as a unit. If the sensing relays expected a two-out-of-six code, they would prevent further computation if more or less than two holes appeared in a given row [**125,** p. 52]. Similar checks were used in nearly every step of a computation. When such a check failed, two results were possible depending upon whether the machine was set for "daytime" use with operating personnel present or for unattended "nighttime" or "weekend" operation. In the first mode, a check failure stopped the machine and sounded an alarm. In the second, a check failure switched the machine immediately to other work [**125,** pp. 63, 64]. In the first case, the check failure was located by operating personnel, whose efforts were facilitated by an elaborate check light panel. However, in the second case the problem simply had to be rerun. This inefficiency, especially the latter, led Hamming to investigate the possibility of automatic error-correction. F. Jessie MacWilliams noted this possibility in [**84,** p. 7].

Hamming did not have priority use of the Model V. In fact, he had only weekend access, and that only when another department wasn't using it. "When they didn't need it, I got it." [**56,** tape 2] Furthermore, for weekend use the machine was placed in the unattended mode, which meant that any errors caused the machine to go on to the next problem. Of this unfortunate circumstance Hamming said:

> Two weekends in a row I came in and found that all my stuff had been dumped and nothing was done. I was really aroused and annoyed because I wanted those answers and two weekends had been lost. And so I said, 'Damn it, if the machine can detect an error, why can't it locate the position of the error and correct it?' [**56,** tape 2].

He then searched for a code that would enable the machine both to detect and to correct errors. His first solution appeared in a memorandum dated July 27, 1947, but was not circulated officially until nearly a year later in a slightly modified form [**49**], [**50**].

The simplest single-error-correcting code given in the 1947 paper was based on an elaboration of the idea of parity checks. The code consists of codewords of length $n = (t+1)^2$, each with $r = t^2$ information digits and $2t + 1$ check digits. To encode the t^2 information digits, place them in a square array and adjoin $2t + 1$ digits on two adjacent sides (including the common corner) so that each row and column in the final square array has an even number of 1's. An example, in the case $t = 2$ is shown below, where we encode the 4-block 1101.

$$1101 \longrightarrow \begin{array}{cc|c} 1 & 1 & \\ 0 & 1 & \\ \hline & & \end{array} \longrightarrow \begin{array}{cc|c} 1 & 1 & 0 \\ 0 & 1 & 1 \\ \hline 1 & 0 & 1 \end{array}$$

The resulting codeword might be written on a single line, as

$$110011101,$$

which is just the sequence of rows in order. The four message digits, instead of being in the first four positions, are in positions 1, 2, 4 and 5. The decoding of a transmitted codeword simply involves placing the codeword in the $(t + 1)$ by $(t + 1)$ array and again performing the row and column parity checks. A single error in transmission will cause the parity checks of the row and column containing that position to fail. The information rate of this code is $t^2/(t + 1)^2$. In case $t = 2$, this is $4/9 = 0.44$, an improvement on the rate $4/12 = 0.33$ of the (12, 4) block repetition code, which has the same number of message digits and is also single-error-correcting.

Hamming noted in the 1948 revision that the corner check symbol could be deleted without losing the error-correction capability of the code. An error in one of the message digits causes two parity checks to fail while an error in a check digit causes only one parity check to fail. For example, suppose

$$11001110 \longrightarrow 01001110.$$

The checking array for the received codeword is

$$\begin{array}{cc|c} 0 & 1 & 0 \\ 0 & 1 & 1 \\ \hline 1 & 0 & \end{array},$$

which shows that both the first row and first column parity checks fail. If

$$11001110 \longrightarrow 11101110,$$

the checking array is

$$\begin{array}{cc|c} 1 & 1 & 1 \\ 0 & 1 & 1 \\ \hline 1 & 0 & \end{array},$$

which shows only the first row parity check failing; the error was, therefore, the right-hand 1 in the top row. The rate of this code is $4/8 = 0.50$. By careful selection of codewords, we have obtained a code with the same rate as the (8, 4) block repetition code but that will also correct single errors. We shall soon show that the (7, 4) Hamming code, first published by Shannon, has the highest rate possible of any single-error-correcting code with four message digits.

Hamming went on in the 1947 memorandum to generalize these "square codes" to "rectangular codes" and then to higher dimensional codes which allowed multiple-error-correction.

Both his 1947 memorandum and its more widely circulated counterpart of June 10, 1948, show that Hamming's main interest was in the practical application of error-correcting codes. They mention both the relay machines and future high-speed electronic machines. For example, in the 1947 paper, Hamming states:

> The purpose of this memorandum is to give some practical codes which may detect and correct all errors of a given probability of occurrence, and which detect errors of even a rarer occurrence. [**49**, p. 1]

Further, in the appendix to the same paper, Hamming remarks:

> In designing high speed digital machines we find that at the speeds usually contemplated so many single operations are per-

formed in an hour as to make the chance of getting the correct answer vanishingly small after an hour's operation.

In the same appendix he suggests that the square single-error-correcting code with $t = 7$ (thus $t^2 = 49$ message digits and $(t + 1)^2 = 64$ total digits) would probably be adequate for the high-speed (electronic) machines,

...we can run about 10α hours [$\alpha = 2300$]—a comfortable figure—before stopping for an error. We will miss an error approximately every $\alpha/2 \times 10^8$ hrs. which is close enough to eternity for most purposes.

In this same vein, Hamming, in the June 10, 1948, memorandum states:

The Relay Computer, operating with a self-checking code, stops whenever an error is detected. Under normal operating conditions this amounts to two or three stops per day. However, if we imagine a comparable electronic computing machine operating at 10^5 times the speed [of the Relay Computer] and with element 10^3 times more reliable than relays, we find two to three hundred stops per day. A self-correcting code, with suitable equipment to effect the corrections, would reduce the stops per day to the vanishing point. In fact, a self-correcting code would enable one to relax the reliability requirements on the individual elements and still have the stops per day vanishingly small. It is hoped that the amount of equipment required to mechanize a self-correcting code would not be excessive, but this question requires further examination. [**50**, pp. 1, 2]

Hamming even tried to extend some of his ideas to known codes, for two paragraphs later in the same memorandum he says:

Because of the wide use of the two-out-of-five-hole codes in the Laboratories, a brief examination has been made of some of the possible modifications of these codes to convert them into self-correcting codes. [**50**, p. 2]

We have not yet come to what are known as the "Hamming Codes," one special case of which Shannon announced in

July, 1948. Hamming did, however, allude to them in his June 10, 1948, memorandum:

> A later memorandum will examine some codes having a minimum redundancy. [**50**, p. 2]

(*Redundancy*, as defined by Hamming, is the reciprocal of the information rate.)

The next memorandum by Hamming on coding is dated September 6, 1948 [**51**]; both this and the one in June circulated among at least ten departments and individuals. In it Hamming pursued single-error-correcting codes almost exclusively, perhaps because he thought that their use would protect future high-speed machines from interruption. Here, too, he used the geometric view and his metric. He wanted an efficient code with r message digits, as few check digits as possible, and yet one that could still correct single errors. This is equivalent to finding a single-error-correcting code with minimum redundancy or maximum information rate. In addition, he wanted each check digit to be a 0 or 1 so that it and the message digits in certain positions would have even parity as a unit, sort of partial parity checks. Finally, upon the reception of such a codeword, each of the parity checks would be calculated in order. The sequence of 0's and 1's thus calculated, called the *checking number* or *syndrome* when written from right to left, and to be distinguished from the set of check digits, would give the position in binary of any single error, with the all-zero word meaning "no error." The $k = n - r$ check digits (which is also the number of digits in the checking number) must be able to describe $r + k + 1 = n + 1$ possibilities since an error could occur in any one of the n places of the codeword or the word might be transmitted without error. Thus

$$2^k \geq n + 1 \tag{1.9}$$

since 2^k is the number of possible checking combinations.

Multiplying both sides of (1.9) by 2^r and then dividing by $n + 1$ gives the fundamental inequality

$$\frac{2^n}{n+1} \geq 2^r$$

from which the minimum n for given r can be determined. Hamming listed Table 1.10 which gives the values of r and k for small n [**51**, p. 4].

TABLE 1.10

n (total)	r (message)	$k = n - r$ (checks)
1	0	1
2	0	2
3	1	2
4	1	3
5	2	3
6	3	3
7	4	3
8	4	4
9	5	4
10	6	4
11	7	4
12	8	4
13	9	4
14	10	4
15	11	4
16	11	5
	etc.	

In particular, this table shows that for $r = 4$, only three check digits are necessary to effect single-error correction. Thus, among codes with four message digits the (8, 4) modified—$(t + 1)^2$ code ($t = 2$, corner element deleted—see pp. 18–19) is not the one with the highest rate. The (7, 4) *Hamming code* has information rate $4/7 = 0.57$. Let us describe this ingenious way of recovering the error position from the three parity digits.

Hamming determined the positions of the various message and check digits within a codeword on the basis of the binary representation of the positive integers. For convenience, Table 1.11 gives these for the first few integers.

TABLE 1.11

Decimal	Binary
1	0001
2	0010
3	0011
4	0100
5	0101
6	0110
7	0111
8	1000
9	1001
10	1010
11	1011
12	1100
13	1101
14	1110
15	1111
16	10000

etc.

Recall that the checking number, which consists of three digits, should locate any single error in a codeword. Its right-hand digit represents the result of the first parity check—a 0 indicates no error while a 1 shows that a single error is in one of the positions of that check. This means, from Table 1.11, that the first parity check is over the positions

$$1, 3, 5, 7, 9, \ldots$$

since the binary representation of these integers has a 1 in the ones position. Similarly, the second parity check is over the positions

$$2, 3, 6, 7, 10, 11, \ldots$$

whose binary representations have a 1 in the twos place. The third parity check is over positions

$$4, 5, 6, 7, 12, 13, 14, 15, \ldots$$

which have a 1 in the fours place, and so on. Instead of placing the check positions at the end of the codeword, Hamming placed the ith check digit in the 2^{i-1}th position. Since 1, 2, 4, 8, ... have but a single 1 in binary, this meant that the check digits would be set independent of one another. In other words, these positions are such that no two check digits check each other. Thus, the setting of the check digits can take place in any order. The message digits were placed in the remaining positions.

To illustrate the encoding process, Hamming chose $n = 7$, the same example Shannon used [**51,** pp. 5, 6], [**54,** p. 152], [**108,** p. 418]. Table 1.10 shows that each codeword has four message digits and three check digits. For example, to encode 1010, place the digits in their appropriate positions—not at 1, 2, or 4:

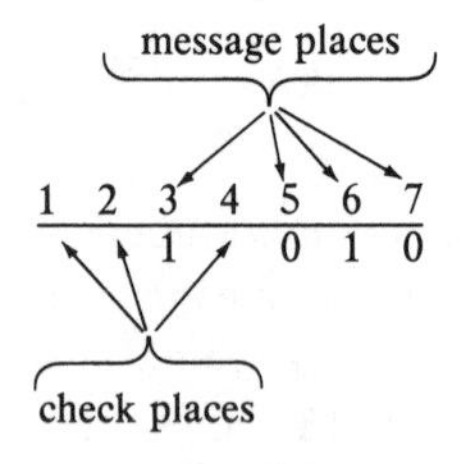

The first parity check is over positions 1, 3, 5 and 7. Hence, the first check digit is $1 + 0 + 0 \equiv 1 \pmod 2$ which will be placed at position "1." Similarly, the second check digit is 1

$+ 1 + 0 \equiv 0 \pmod 2$ and the third is $0 + 1 + 0 \equiv 1 \pmod 2$. The resulting codeword is

$$\underline{1}\underline{0}1\underline{1}010.$$

The decoding process is just as easy. If no error in transmission occurs, the checking number is, of course, 000. However, suppose

$$1011010 \longrightarrow 1011000.$$

The parity checks are

$$1 + 1 + 0 + 0 \equiv 0,$$

for positions 1, 3, 5, 7,

$$1 + 0 + 0 + 0 \equiv 1,$$

for positions 2, 3, 6, 7 and

$$1 + 0 + 0 + 0 \equiv 1,$$

for positions 4, 5, 6, 7. The checking number is

$$110$$

which indicates that position 6 is in error. The process is the same wherever the error occurs.

Is this the best possible single-error-correcting code with four message digits, i.e., does it have the highest information rate? To show that it is the best, first note that since the minimum distance of any such code must be three, Hamming 1-spheres centered on the codewords must be disjoint. Now, the Hamming 1-sphere of any codeword in this case ($n = 7$) contains eight vertices. All told, the sixteen Hamming 1-spheres in this example contain a total of $16 \cdot (1 + 7) = 2^7$ (hence all) vertices of the unit 7-cube. Not only is this code the best possible for those with four message digits, but it is perfect.

Although Hamming generalized this code to all dimensions, only the $(n, r) = (2^k - 1, 2^k - 1 - k)$ Hamming codes are perfect. To show this, note that it is necessary that the number of vertices in a Hamming 1-sphere, centered on a codeword vertex, be a power of two. This means that

$$\binom{n}{0} + \binom{n}{1} = 1 + n = 2^m$$

where m is an integer. This implies that

$$n = 2^m - 1.$$

From Table 1.10, $m = k$ so that the condition is also sufficient.

All the codes discussed so far have had codewords with distinct message and check digits. In the September, 1948, memorandum, Hamming called codes with this property, *systematic*. He also briefly discussed nonsystematic codes and gave an example for $n = 6$, under the heading "Some Remarks on How Not to Construct a Code," which had six codewords—two fewer than his (6, 3) code [**51**, p. 10]. But he did not find them of practical interest:

> There are, furthermore, codes which do not use the systematic checks we have described; they seem, however, to be more difficult to use in practice. [**51**, p. 7]

All the coding theory described in Hamming's 1950 paper [**54**] had appeared over a year and one-half earlier in the interdepartmental memorandum of September 1948 [**51**] and had been alluded to even earlier in the memorandum of June 1948 [**50**]. Why didn't Hamming publish his work sooner? Why did he wait so long after Shannon had published the (7, 4) example? Hamming explains that this delay was due to patent considerations,

> I wrote the paper [and sent it] to the *B.S.T.J.* [*Bell System Tech. J.*] to try and get it out in print and that is what hung it up, bang, like that. The Patent Department [Legal and Patent Division]

> would not release the thing until they had patent coverage. [**56**, tape 4]

Hamming was skeptical about securing a patent.

> I didn't believe that you could patent a bunch of mathematical formulas. I said they couldn't. They said, 'Watch us.' They were right. And since then I have known that I have a very weak understanding of patent laws because, regularly, things that you shouldn't be able to patent—it's outrageous—you can patent. [**56**, tape 2]

Because of the decision of the Legal and Patent Division, much of Hamming's work prior to the publication of his 1950 paper is available.

> When it became obvious that it was going to be an important patent and my end of it was all done, I knew enough to grab all the things for patent reasons—just threw them all in an envelope. [**56**, tape 2]

Included in this collection are preliminary drafts of some of the memoranda, scratch work, several drafts of the 1950 paper and other related material. Much of the preliminary material is undated.

Section 3. The Hamming-Holbrook Patent

In order to patent the code—in particular the (7, 4) code—the Legal and Patent Division needed diagrams and descriptions of the switching circuitry necessary to implement the code in addition to Hamming's mathematical work. Bernard D. Holbrook, of the Switching Research Department at the Laboratory, was asked to draw up the appropriate electric circuitry. Hamming had known Holbrook for some time:

> I was a pure mathematician—I felt somewhat lost at the place. Every once in a while I got terribly discouraged at the department being mostly electrical engineering. I had bumped into Holbrook and I used to occasionally, when my morale was low, go down and sit in Holbrook's office and just talk to him. [**56**, tape 4]

Holbrook had also received both the June and September, 1948, memoranda [**50,** cover sheet], [**51,** cover sheet]. Hamming suggests that he had talked to Holbrook concerning the codes even before Holbrook was asked to work on the circuitry.

> He probably knew about it because I was sitting on his desk complaining sometime about 'My God, I don't belong in this place at all.' [**56,** tape 4]

Holbrook confirms this in saying that it was he, in fact, who instigated the patent process:

> I looked at it and said, 'We ought to do something about getting a patent on this.' So I went to my boss and he said, 'Well, you can draw up a circuit, do it.' So I did. [**64**A]

By October 19, 1948, Holbrook had already started designing the circuitry [**64,** sketches 1, 2]. As his work progressed, each page was stamped with the words "witnessed and understood" accompanied by the signatures of Hamming, Holbrook and A. E. Joel and the date [**64,** sketches]. Though Holbrook finished his work about a month later, it was not marked "received" by the Legal and Patent Division until February 16, 1949 [**63**], [**64,** back of sketch 1]. In the meantime, on February 1, the Legal and Patent Division had received copies of the June and September, 1948, Hamming memoranda, as the Division was not on the original circulation list [**50**], [**51**] (cover sheets). Holbrook submitted an additional memorandum to the Legal and Patent Division dated August 9, 1949 [**64**]. One reason for the numerous dates being listed is that a little later we will weave the work of Golay into the same time framework.

Holbrook was not just interested in designing the circuitry:

> Since I had the circuit, I turned it over to a technical assistant and said, 'Let's build something so that we can demonstrate this.' . . . I had this thing built essentially for two reasons. One of them was that I thought that if we were getting a patent, we ought to tell it to the switching development people in the Laboratories and

> this would be an easy way to do it. The second was that I wanted to see how the damn thing would work when we put it on relays. [**64**A]

As of April 12, 1949, Hamming had written the final form of the paper which was to appear, with minor changes, in April, 1950, in the Bell System Technical Journal. The latter paper, accepted for publication on February 7, 1950, is the one noted in references on Hamming's work in error-correcting codes [**103**]. Two distinct versions of the April 12 paper exist, one of which was received by the Legal and Patent Division on May 24, 1949 [**52**], [**53**].

The Hamming-Holbrook patent was applied for on January 11, 1950, and issued on May 15, 1951 [**60**]. The reception of Hamming's contribution and the working model of Holbrook's was less than enthusiastic. As Hamming was to remark some twenty-seven years later,

> It [a demonstration of the working model] was to be given every hour or half-hour to various groups. So Holbrook gave the talks—for he was very good at explaining—and I was there in case any question ever arose. None did. I would hear people coming out of the demonstration saying, 'It can't be done.' . . . They were free to ask questions, they were free to come up and play with the equipment . . . but they wouldn't raise their hands to say a word. They would make no noises and simply walk out. I was stunned to realize what the old-line engineer was. They are not the ones now. No, we're talking about the ones who were there in '49 and '50 and who were then old—who had been there since, say, 1910. [**56,** tape 2]

As it turned out, the patent soon became available for royalty free licenses to everyone in 1956 as the result of a settlement of an antitrust case against the Bell System [**35A**].

Section 4. The Hamming Codes Are Linear

There is more mathematical structure in the Hamming codes than we have mentioned. In fact, they form a group as Hamming recalled in 1977,

> I knew that the sum of two symbols was a symbol. [57]

However, there is no reference in either the 1950 B.S.T.J. article or any earlier memorandum to this algebraic property. Nevertheless, this observation implies that the (n, r) Hamming code is a subgroup of the group of all binary n-tuples under vector addition modulo 2. Indeed, it can be viewed as a vector subspace of $[GF(2)]^n$ over the field GF(2). Hamming went on to add:

> I certainly knew in a vague way the group structure, but I did not exploit that. [57]

Any code which can be viewed as a subspace of some vector space is called *linear.* Let us see why (n, r) Hamming codes are linear.

Take, for example, the (7, 4) code. Recall that each received word was subjected to a system of three parity checks which together produced a 3-digit checking number.

The checking number 000 meant that the received word was a codeword; only in this 000 case is the received word a codeword. Thus, it is reasonable to suspect that the (7, 4) code is the kernel of some linear transformation from $[GF(2)]^7$ to $[GF(2)]^3$, hence, a group. To determine this linear transformation, let H be the 3×7 matrix

$$\begin{bmatrix} 0 & 0 & 0 & 1 & 1 & 1 & 1 \\ 0 & 1 & 1 & 0 & 0 & 1 & 1 \\ 1 & 0 & 1 & 0 & 1 & 0 & 1 \end{bmatrix} \tag{1.12}$$

whose columns are the binary numbers from one to seven. The matrix H determines a linear transformation from $[GF(2)]^7$ to $[GF(2)]^3$. For example, let W be the codeword given by the 1×7 matrix

$$[1 \quad 0 \quad \underline{1} \quad 1 \quad \underline{0} \quad \underline{1} \quad \underline{0}];$$

this is the codeword formed without error from the 4-block 1010 whose positions have been underlined (see pp.24–25).

Then

$$HW^T = \begin{bmatrix} 0 & 0 & 0 & 1 & 1 & 1 & 1 \\ 0 & 1 & 1 & 0 & 0 & 1 & 1 \\ 1 & 0 & 1 & 0 & 1 & 0 & 1 \end{bmatrix} \begin{bmatrix} 1 \\ 0 \\ 1 \\ 1 \\ 0 \\ 1 \\ 0 \end{bmatrix} = \begin{bmatrix} 0 \\ 0 \\ 0 \end{bmatrix}$$

where all multiplication is modulo 2. Note that 1's in the bottom row of H occupy the positions of the first parity check while those in the second and top rows are in the positions of the second and third parity checks, respectively. This means that the matrix multiplication amounts to computing the three parity checks. The product, when read from top to bottom, is thus the checking number.

For instance, imagine that

$$[1011010] \longrightarrow [1011000].$$

The matrix HW^T in this case is

$$\begin{bmatrix} 0 & 0 & 0 & 1 & 1 & 1 & 1 \\ 0 & 1 & 1 & 0 & 0 & 1 & 1 \\ 1 & 0 & 1 & 0 & 1 & 0 & 1 \end{bmatrix} \begin{bmatrix} 1 \\ 0 \\ 1 \\ 1 \\ 0 \\ 0 \\ 0 \end{bmatrix} = \begin{bmatrix} 1 \\ 1 \\ 0 \end{bmatrix}$$

which shows that the sixth position in [1011000] is erroneous.

The form of columns, 1, 2 and 4 of H indicate that H has rank three and thus its kernel has dimension four. The kernel of the linear transformation defined by H, therefore, has $2^4 = 16$ elements. Since any codeword, when operated on by H, yields

$$\begin{bmatrix} 0 \\ 0 \\ 0 \end{bmatrix},$$

the kernel of H is the (7, 4) Hamming code. Thus, we have a purely algebraic description of the code as the kernel of a certain linear map. Similarly, the (n, r) Hamming code is a vector space of dimension r over the field GF(2).

Whenever a code can be represented as the kernel of some linear transformation whose matrix is H, the matrix H is called the *parity check matrix* for that code, although early usage of this term was slightly different [**35,** Fig. 5]. The column vector HW^T is the *syndrome* of W. As already noted, if no more than one error occurs, the syndrome indicates in binary the location within the codeword of the error for the Hamming codes.

Hamming came close to discovering the matrix H when he used the matrix

$$\begin{bmatrix} 1 & 0 & 1 & 0 & 1 & 0 & \cdots \\ 0 & 1 & 1 & 0 & 0 & 1 & \cdots \\ 0 & 0 & 0 & 1 & 1 & 1 & \cdots \\ 0 & 0 & 0 & 0 & 0 & 0 & \cdots \\ \vdots & \vdots & \vdots & \vdots & \vdots & \vdots & \end{bmatrix}$$

in showing the equivalence of his two definitions of a systematic code [**54,** p. 157].

The observation that the (n, r) Hamming code is a group suggests the following simple proof that the minimum distance between any two of its codewords is at least three. First, define the (Hamming) *weight*, w, of a codeword as the number of its nonzero digits. For example,

$$w(1011010) = 4.$$

It should be noted that although Hamming did not introduce this notion, it is just the distance of the given codeword from the origin. Irving S. Reed introduced a notion, equivalent in the binary case, in 1954 [**102,** p. 38].

The distance between two codewords is simply the weight of their algebraic difference. Thus, showing that the minimum distance between any two codewords of a linear code is at least three is equivalent to showing that the weight of any nonzero codeword is at least three. If D is any codeword of the (n, r) Hamming code of weight one or two, i.e., D has one or two 1's, then the equation

$$HD^T = \begin{bmatrix} 0 \\ \cdot \\ \cdot \\ \cdot \\ 0 \end{bmatrix}$$

means that either one column or the vector sum of two columns of H is the zero column. Since all the columns of H are distinct and nonzero, this is impossible. Thus, the minimum weight of any nonzero codeword is three. In particular, this shows that the (n, r) Hamming codes are single-error-correcting.

Section 5. The Work of Golay

Another name clearly identified with the origins of coding theory is Marcel J. E. Golay. It is his work that links coding theory to Leech's results in sphere packing.

While working at the Signal Corps Engineering Laboratories at Fort Monmouth, New Jersey, Golay read Shannon's account of the (7, 4) Hamming code soon after it was published in July, 1948; and as he said in a letter in 1977,

> ...generalized it fairly expeditiously to perfect one-error-correcting codes based on any prime number, and then became curious about the possibility of perfect multi-error-correcting codes. [**40**], [**44**]

His report, accepted February 23, 1949, appeared in June in the *Proceedings of the I.R.E.* (*I.E.E.E.*). The paper was titled "Notes on Digital Coding," and occupied a little more than half a page in the Correspondence section [**40**]. E. R. Berlekamp has called it the "best single published page" in coding theory [**13,** p. 4].

Golay was born and educated in Switzerland which he left in 1924 with the License in Electrical Engineering degree from the Federal Institute of Technology in Zurich [**67**]. Although his formal training did not include much mathematics, his interest in number theory was aroused early:

> At the age of 15 I was a member of a little club in my hometown where most of the others were bigger boys and some of them were mathematicians. I remember one in particular of whom it was fun to ask questions. They took pleasure in tutoring a little bit. [**46**]

He did not stop there.

> At the age of 18 there were nice occasions in Zurich where I was to listen to talks by higher mathematicians like Kollross and Weyl and people like that. I didn't get too much from them, but anyway they were stimulating. [**46**]

However, he had no formal courses in pure mathematics. After working at the Bell Telephone Laboratories in the Cable Department in Chicago (Hawthorne), Illinois, from 1924 to 1928, he began graduate work at the University of Chicago where he obtained a Ph.D. in physics in 1931. In the same year he had started to work for the Signal Corps Engineering Laboratories [**46**], [**67**].

Golay became acquainted with a small part of Hamming's work through the following description of the (7, 4) Hamming code given by Shannon:

> Let a block of seven [binary] symbols be $X_1, X_2, \ldots, X_7$. Of these X_3, X_5, X_6 and X_7 are the message symbols and chosen arbitrarily by the source. The other three are redundant and calculated as follows:
>
> X_4 is chosen to make $\alpha = X_4 + X_5 + X_6 + X_7$ even
> X_2 is chosen to make $\beta = X_2 + X_3 + X_6 + X_7$ even
> X_1 is chosen to make $\gamma = X_1 + X_3 + X_5 + X_7$ even.
>
> When a block of seven is received, α, β and γ are calculated and if even, called zero, if odd, called one. The binary number $\alpha\beta\gamma$ then gives the subscript of the X_i that is incorrect (if 0, there was no error). [**108**, p. 418]

This code had been presented as an example of one whose rate exactly matched the capacity of a certain noisy channel. The channel, which he acknowledged as "somewhat unrealistic," was to consist of 7-blocks of binary digits. Noise was to affect each 7-block so that it was either transmitted without error or exactly one digit would be in error. The channel appears in Figure 1.13.

Each 7-block on either the input side or output side has eight attachments with all probabilities chosen to be 1/8. The channel capacity is thus

$$\log_2 128 + 8\cdot\frac{1}{8}\log_2\frac{1}{8}$$

$$= 7 - \log_2 8$$

$$= 4 \text{ bits per 7-block.}$$

This amounts to 4/7 bits per digit [**108,** pp. 416–418]. Recall that this is exactly the rate of the (7, 4) Hamming code.

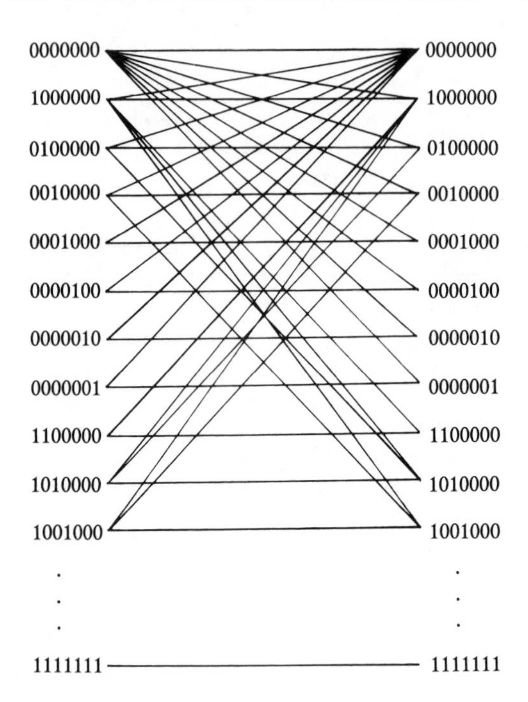

FIG. 1.13. Shannon's channel for the (7, 4) Hamming code.

Shannon's example inspired Golay to fashion perfect single-error-correcting codes based on p symbols for any prime p.

> I had been thinking about information theory for quite a while, when I was involved with radar. I said, 'What do we want out of radar? After all, we want information.' And I began to acquire the idea that a bit of information cost about one $kT \log_e 2$. Then, later on, I pinned it down a little more. But I didn't come to—I didn't think of—the idea of Hamming, of coding or like that. But when I read the [Shannon] paper, it was the key because I was ripe for these progresses. I had already thought about it. [**46**]

In his paper he first noted that the block length n of a perfect code for a given prime p must be of the form

$$n = \frac{p^k - 1}{p - 1}. \tag{1.14}$$

This equation is easily derived from the fact that each (Hamming) sphere of radius one contains $1 + n(p-1)$ points. For a code to be perfect, this number must divide p^n and hence must be of the form p^k for some integer k. Consequently, $1 + n(p-1) = p^k$, which is equivalent to Equation (1.14). Golay then showed how to generate such a code with each codeword having k check digits. The check digits X_1, $\ldots$, X_k were determined from the message digits $Y_1, \ldots,$ Y_r, $r = n - k$, by the congruences of the form:

$$E_i \equiv X_i + \sum_{j=1}^{r} a_{ij} Y_j \equiv 0 \pmod{p}, \tag{1.15}$$

where the k by n matrix $[a_{ij}]$ is constructed by induction on k. In the decoding process, each of the E_i's is calculated and $E_i \cdots E_k$ is the base p representation of a number which locates any single error and indicates the correction. Golay did all this knowing only the brief result of Hamming given by Shannon. He did not illustrate the process until 1958, and then with a slight correction [**42,** pp. 103, 104].

Let us examine the construction for the case $p = 2$. If $k = 1$, then $n = 1$ and the code consists of a single binary check symbol X_1. The coding matrix $[a_{ij}]$ consists of a single "1" and since there are no message symbols, $X_1 = 0$ according to the congruence relation (1.15). In the passage from k to $k + 1$, the matrix $[a_{ij}]$ with k rows and n columns was to be repeated p times horizontally, a $(k+1)$th row added consisting of n 0's followed by n 1's, etc., up to n $(p-1)$'s and an additional column consisting of k 0's followed by one 1. The change he made in 1958 was to reverse the order of the symbols in the added row (otherwise the matrix would not have the parity checks X_i neatly assembled on the right). In general the matrix $[a_{ij}]$ would take the form of this k by $k + r$ matrix:

$$\begin{bmatrix} a_{11} & a_{12} & \cdots & a_{1r} & 1 & 0 & \cdots & 0 \\ a_{21} & a_{22} & \cdots & a_{2r} & 0 & 1 & \cdots & 0 \\ \vdots & \vdots & & \vdots & \vdots & \vdots & & \vdots \\ a_{k1} & a_{k2} & & a_{kr} & 0 & 0 & & 1 \end{bmatrix}.$$

The congruence relations in (1.15) are easily expressed in a single matrix equation:

$$\begin{bmatrix} E_1 \\ \vdots \\ E_k \end{bmatrix} \equiv [a_{ij}] \begin{bmatrix} Y_1 \\ \vdots \\ Y_r \\ X_1 \\ \vdots \\ X_k \end{bmatrix} \equiv \begin{bmatrix} 0 \\ \vdots \\ 0 \end{bmatrix} \pmod{p}.$$

For example, the matrix for $k = 2$ is

$$\begin{bmatrix} 1 & 1 & 0 \\ 1 & 0 & 1 \end{bmatrix}$$

while that for $k = 3$ is

$$\begin{bmatrix} 1 & 1 & 0 & 1 & 1 & 0 & 0 \\ 1 & 0 & 1 & 1 & 0 & 1 & 0 \\ 1 & 1 & 1 & 0 & 0 & 0 & 1 \end{bmatrix}. \tag{1.16}$$

This latter matrix is the parity check matrix for the (7, 4) Hamming code with several column interchanges (compare with matrix 1.12). The code is the kernel of this matrix when the matrix is viewed as a linear transformation from the vector space $[GF(2)]^7$, which has 2^7 elements, to $[GF(2)]^3$, which has only 2^3 elements. The codewords in the kernel of this matrix differ from those of the Hamming code only in the location of the message and check digits. For example, note that the codeword

$$1110000$$

is in the kernel of matrix 1.12 but not in the kernel of matrix 1.16. Two such codes, differing only in the order of their entries, Hamming considered *equivalent* [**54,** p. 156]. If a single error occurs in transmission, the matrix

$$\begin{bmatrix} E_1 \\ \cdot \\ \cdot \\ \cdot \\ E_k \end{bmatrix}$$

is simply the column of $[a_{ij}]$ whose number is that of the position of the error in the codeword. Note that the number $E_1 \ldots E_k$ is not the position number of the error. For example, suppose

$$1010100 \longrightarrow 1010110.$$

Then

$$\begin{bmatrix} E_1 \\ E_2 \\ E_3 \end{bmatrix} \equiv \begin{bmatrix} 1 & 1 & 0 & 1 & 1 & 0 & 0 \\ 1 & 0 & 1 & 1 & 0 & 1 & 0 \\ 1 & 1 & 1 & 0 & 0 & 0 & 1 \end{bmatrix} \begin{bmatrix} 1 \\ 0 \\ 1 \\ 0 \\ 1 \\ 1 \\ 0 \end{bmatrix} \equiv \begin{bmatrix} 0 \\ 1 \\ 0 \end{bmatrix}$$

which indicates that the sixth position is in error. In the binary case the correction is obvious.

The prime $p = 3$ provides a more interesting example. When $k = 1$, $[a_{ij}]$ is again the matrix [1]. When $k = 2$, $[a_{ij}]$ becomes

$$\begin{bmatrix} 1 & 1 & 1 & 0 \\ 2 & 1 & 0 & 1 \end{bmatrix}.$$

For $k = 3$, $[a_{ij}]$ is thus

$$\begin{bmatrix} 1 & 1 & 1 & 0 & 1 & 1 & 1 & 0 & 1 & 1 & 1 & 0 & 0 \\ 2 & 1 & 0 & 1 & 2 & 1 & 0 & 1 & 2 & 1 & 0 & 1 & 0 \\ 2 & 2 & 2 & 2 & 1 & 1 & 1 & 1 & 0 & 0 & 0 & 0 & 1 \end{bmatrix}.$$

For the sake of simplicity, the correction process will be illustrated when $k = 2$. The nine codewords are:

$$\begin{matrix} 0000 & 1021 & 2012 \\ 0122 & 1110 & 2101 \\ 0211 & 1202 & 2220 \end{matrix}$$

Suppose that

$$1202 \longrightarrow 2202.$$

Then

$$\begin{bmatrix} E_1 \\ E_2 \end{bmatrix} \equiv \begin{bmatrix} 1 & 1 & 1 & 0 \\ 2 & 1 & 0 & 1 \end{bmatrix} \begin{bmatrix} 2 \\ 2 \\ 0 \\ 2 \end{bmatrix} \equiv \begin{bmatrix} 1 \\ 2 \end{bmatrix} \equiv 1 \begin{bmatrix} 1 \\ 2 \end{bmatrix}$$

indicates that the digit in the first position is in error and the correction is to subtract 1 modulo 3 from the digit in the codeword in that position. If, instead,

$$1202 \longrightarrow 0202,$$

then

$$\begin{bmatrix} E_1 \\ E_2 \end{bmatrix} \equiv \begin{bmatrix} 1 & 1 & 1 & 0 \\ 2 & 1 & 0 & 1 \end{bmatrix} \begin{bmatrix} 0 \\ 2 \\ 0 \\ 2 \end{bmatrix} \equiv \begin{bmatrix} 2 \\ 1 \end{bmatrix} \equiv 2 \begin{bmatrix} 1 \\ 2 \end{bmatrix}$$

which indicates that the digit in the first position is again in error and the correction is to subtract, modulo 3, a 2 from that digit. Single-error correction is always possible since no column of $[a_{ij}]$ is zero and any two columns are linearly independent over GF(3). Note that $[a_{ij}]$ consists of all possible nonzero columns of two elements over GF(3) whose first nonzero entry is a 1.

The process described easily generalizes for any prime p and positive integer k. Although the existence of perfect single-error-correcting codes using p^m symbols was known before 1958, a method for constructing them was not.

S. K. Zaremba, in 1952 [**128,** p. 243], published the following theorem concerning finite abelian groups:

> *Let G be an Abelian group with n base elements, $g_1, \ldots, g_n$, each of order p^m, where p is a prime, and let S denote the set of $\mu = n(p^m - 1) + 1$ elements $1, g_1, g_1^2, \ldots, g_1^{p^m-1}, \ldots, g_n, g_n^2, \ldots, g_n^{p^m-1}$ of G. If μ is a power of p^m, then there exists a subset H of G composed of p^{mn}/μ elements and satisfying G = HS.*

A word on the necessity of the condition is in order. Clearly, if such an H exists, $n(p^m - 1) + 1$ would divide p^{mn}, hence equal p^t for some integer t. Thus,

$$n(p^m - 1) = p^t - 1$$

so that

$$p^t - 1 \equiv 0 \,(\mathrm{mod}(p^m - 1)).$$

Write $t = qm + r$, $0 \leq r < m$. Then

$$p^{qm}p^r \equiv 1 \,(\mathrm{mod}(p^m - 1)).$$

But

$$p^m \equiv 1 \,(\mathrm{mod}(p^m - 1))$$

so that

$$p^r \equiv 1 \,(\mathrm{mod}(p^m - 1))$$

whence

$$r = 0.$$

To show how Zaremba's theorem implies the existence of perfect single-error-correcting codes of length n, write the group elements as n-tuples. Thus,

$$1 = (1, 1, \ldots, 1)$$

$$g_i = (1, \ldots, 1, \underbrace{g}_{i\text{th position}}, 1, \ldots, 1)$$

and

$$g_i^j = (1, \ldots, 1, g^j, 1, \ldots, 1).$$

Let $h = (g^{h_1}, g^{h_2}, \ldots, g^{h_n})$ be an element of H. A change in the ith coordinate is equivalent to multiplying h by g_i^j for some j:

$$\begin{aligned} hg_i^j &= (g^{h_1}, g^{h_2}, \ldots, g^{h_i}g^j, \ldots, g^{h_n}) \\ &= (g^{h_1}, g^{h_2}, \ldots, g^{h_i+j}, \ldots, g^{h_n}). \end{aligned}$$

This new element is in Hg_i^j and in no other such coset since the equations $|G| = p^{mn}$ and $|H| = p^{mn}/\mu$ imply that all such cosets are pairwise disjoint. If the elements of H are considered as the codewords of a code, it can then be seen that any single error is correctable.

Observe that the set $G_1 \times \cdots \times G_n$, where G_i is the subgroup of G generated by g_i, can be interpreted as an n-dimensional cube of side p^m composed of p^{mn} unit cubes. The perfect packing corresponding to the set S amounts to placing rooks on the cells of this "chessboard" in such a way that each cube is attacked or occupied by exactly one rook. Some twenty-two years later, S. K. Stein in **[117]** was to explore this type of packing to construct tilings of Euclidean space by certain star bodies that are denser than any lattice tiling by them.

Golay met Zaremba in London in the middle 1950's. Reminiscing about this meeting, Golay recalled that:

> Zaremba showed quite a bit of interest in the [1949] paper. He said, 'Well, Golay, you have said more in half a page than others have said in twenty pages later on.' **[46]**

This encounter took place sometime between 1954 and 1957, since it had to occur after Golay's 1954 paper was written, which mentions only that the perfect codes with p symbols are known to exist, and, as Golay recalls, it occurred a year or so prior to the submitting of his 1958 paper (February 25, 1958). This meeting inspired a large part of Golay's 1958 paper **[46]**.

In that 1958 paper Golay stated the necessary condition for the existence of perfect codes of length n using p^m symbols:

$$n = \frac{(p^m)^k - 1}{p^m - 1}$$

for some integer k. He succeeded in constructing single-error-correcting codes of length n for $m = 2$ and for some primes p with m greater than 2 **[42]**. He also credits Zaremba with the solution in case $p^m = 2^2$ but was not content with his own generalizations [**46**]. The matter was completely solved, including some questions Golay had raised, by John Cocke in 1959 who in one page gave the construction of such codes over GF(p^m) for all primes p and all integers $m \geq 2$ [**22**].

In retrospect, had Golay known about finite fields in 1949 (which he did not [**46**]), he could have applied the method that he used in the prime case directly to the prime power case. For example, let $p = m = k = 2$. Then

$$n = \frac{(p^m)^k - 1}{p^m - 1} = \frac{2^4 - 1}{2^2 - 1} = 5.$$

Let the field elements of GF(2^2) be $0, 1, x, x+1$, with all multiplications carried out modulo the polynomial $x^2 + x + 1$ which, in the ring of polynomials with coefficients in GF(2), is irreducible. When $k = 1$, the matrix $[a_{ij}]$ is again [1]. Thus, for $k = 2$, $[a_{ij}]$ becomes

$$\begin{bmatrix} 1 & 1 & 1 & 1 & 0 \\ x+1 & x & 1 & 0 & 1 \end{bmatrix}.$$

Suppose that the codeword

$$x \quad 0 \quad x \quad 0 \quad x+1 \longrightarrow x \quad x \quad x \quad 0 \quad x+1.$$

The matrix calculation

$$\begin{bmatrix} E_1 \\ E_2 \end{bmatrix} \equiv \begin{bmatrix} 1 & 1 & 1 & 1 & 0 \\ x+1 & x & 1 & 0 & 1 \end{bmatrix} \begin{bmatrix} x \\ x \\ x \\ 0 \\ x+1 \end{bmatrix} \equiv \begin{bmatrix} x \\ x+1 \end{bmatrix}$$

$$\equiv x \begin{bmatrix} 1 \\ x \end{bmatrix}$$

indicates that the element in the second position is in error and the received codeword may be corrected by subtracting x from that element.

In general, the matrix $[a_{ij}]$ consists of all nonzero columns of k elements from the field having 1 as the leading nonzero element. Each code is linear since it is the kernel of a linear transformation. Since no column of $[a_{ij}]$ is zero and any two are linearly independent over the field $\mathrm{GF}(p^k)$, the minimum weight of any nonzero codeword is three. Consequently, the codes are single-error-correcting.

It should be noted that the additive structure of $\mathrm{GF}(p^k)$ is not a cyclic group when $k > 1$. Thus, this generalization of Golay's 1949 work is not the (nongroup code) subset H of the Abelian group G in Zaremba's theorem. In fact, Zaremba showed that H cannot be a subgroup of G if $k > 1$ [**128,** pp. 243, 244]. We also should note that nonbinary codes have not made nearly the impact in practical use as the binary ones have since most electronic equipment operates in a binary mode.

Golay's 1949 paper did not conclude with the perfect single-error-correcting codes. He mentioned the trivial two-word binary codes of length $2k + 1$ which are k-error correcting—the code consisting of the all-zero word and the all-one word—and then gave matrices generating two other multi-error-correcting codes, now called the *Golay codes*. These latter codes have had far-reaching implications for sphere packing and simple groups, some of which we will trace in Chapters 2 and 3.

Golay first stated the necessary condition for the existence of a perfect binary code which could correct more than one error:

> . . . the existence of three or more first numbers of a line of Pascal's triangle which add up to an exact power of 2. [**40**]

This is the same as saying that the number of vertices of the unit n-cube contained in a (Hamming) sphere of radius $R \geq$

3 centered on a vertex must divide the total number of vertices of the cube. Golay then noted that the two following such sums of binomial coefficients coincide with a power of 2:

$$\sum_{i=0}^{2} \binom{90}{i} = 2^{12} \quad \text{and} \quad \sum_{i=0}^{3} \binom{23}{i} = 2^{11}.$$

Using a counting argument, he showed that no systematic (separate message and check digits) perfect double-error-correcting (90, 78) code exists. Golay credits Zaremba in 1958 for showing the nonexistence of a nonsystematic (90, 78) code, thereby completely solving that case [**42,** p. 108].

The second equation, $\Sigma_{i=0}^{3}\binom{23}{i} = 2^{11}$ suggests the possibility for the existence of a perfect 3-error-correcting (23, 12) code. The matrix which Golay gave for it in 1949 is presented in Figure 1.17.

	Y_1	Y_2	Y_3	Y_4	Y_5	Y_6	Y_7	Y_8	Y_9	Y_{10}	Y_{11}	Y_{12}
X_1	1	0	0	1	1	1	0	0	0	1	1	1
X_2	1	0	1	0	1	1	0	1	1	0	0	1
X_3	1	0	1	1	0	1	1	0	1	0	1	0
X_4	1	0	1	1	1	0	1	1	0	1	0	0
X_5	1	1	0	0	1	1	1	0	1	1	0	0
X_6	1	1	0	1	0	1	1	1	0	0	0	1
X_7	1	1	0	1	1	0	0	1	1	0	1	0
X_8	1	1	1	0	0	1	0	1	0	1	1	0
X_9	1	1	1	0	1	0	1	0	0	0	1	1
X_{10}	1	1	1	1	0	0	0	0	1	1	0	1
X_{11}	0	1	1	1	1	1	1	1	1	1	1	1

FIG. 1.17

He did not explain either how he constructed it or why the code it generates is 3-error-correcting. Recall the relation (1.15) that the X_i's and Y_j's must satisfy for the (23, 12) code:

$$X_i + \sum_{j=1}^{12} a_{ij} Y_j \equiv 0 \pmod 2, \quad i = 1, \ldots, 11.$$

Since the code is binary, these congruences may be rewritten as

$$X_i \equiv \sum_{j=1}^{12} a_{ij} Y_j \pmod 2, \quad i = 1, \ldots, 11$$

or

$$\begin{bmatrix} X_1 \\ \cdot \\ \cdot \\ \cdot \\ X_{11} \end{bmatrix} \equiv [a_{ij}] \begin{bmatrix} Y_1 \\ \cdot \\ \cdot \\ \cdot \\ Y_{12} \end{bmatrix} \pmod 2.$$

For example,

100000000000

is encoded as

100000000000 ¦ 11111111110.

By attaching I_{11} to the matrix $[a_{ij}]$, we obtain a matrix of the same form as that of the matrices for the single-error-correcting codes. This new matrix represents a linear transformation from the vector space $[GF(2)]^{23}$ with 2^{23} elements to $[GF(2)]^{11}$ with 2^{11} elements. The (23, 12) Golay code is then the kernel of this linear transformation.

Since the code is linear, showing that it is 3-error-correcting is the same as showing that the nonzero codewords have at least seven 1's. This, in turn, can be established by proving that the sum modulo 2 of six or fewer columns of the matrix

$$\begin{bmatrix} a_{1,1} & \cdots & a_{1,12} & 1 & 0 & \cdots & 0 \\ \vdots & & \vdots & 0 & 1 & & \vdots \\ \vdots & & \vdots & \vdots & & \ddots & \vdots \\ a_{11,1} & \cdots & a_{11,12} & 0 & \cdots & & 1 \end{bmatrix}$$

is never the zero vector (since the product of this matrix with any vector in $[GF(2)]^{23}$ is simply the sum modulo 2 of the columns of this matrix in the positions where the vector has 1's). In view of the structure of the matrix, this quickly reduces to the following checks on the first twelve columns:

The sum of any six columns is not zero.
The sum of any five columns has at least two 1's.
The sum of any four columns has at least three 1's.
The sum of any three columns has at least four 1's.
The sum of any two columns has at least five 1's.
Any column has at least six 1's.

(1.18)

When asked if there was some easy way to verify these conditions, Golay replied that "...it is something which can be verified by inspection." [45] There are

$$\sum_{i=1}^{6} \binom{12}{i} = 2509$$

equations to be "verified by inspection." Soon we shall show how some of these calculations can be simplified.

Golay's 1954 paper [**41,** p. 26] gives the following geometric rationale for determining the source of the first ten symbols in each of the last eleven columns of his matrix $[a_{ij}]$ in Figure 1.17 which we shall now examine.

Consider five lines in the plane, no two of which are parallel and no three of which are concurrent. Label these lines A,

B, C, D and E and their ten respective intersections by AB, AC, AD, etc. This is illustrated in Figure 1.19.

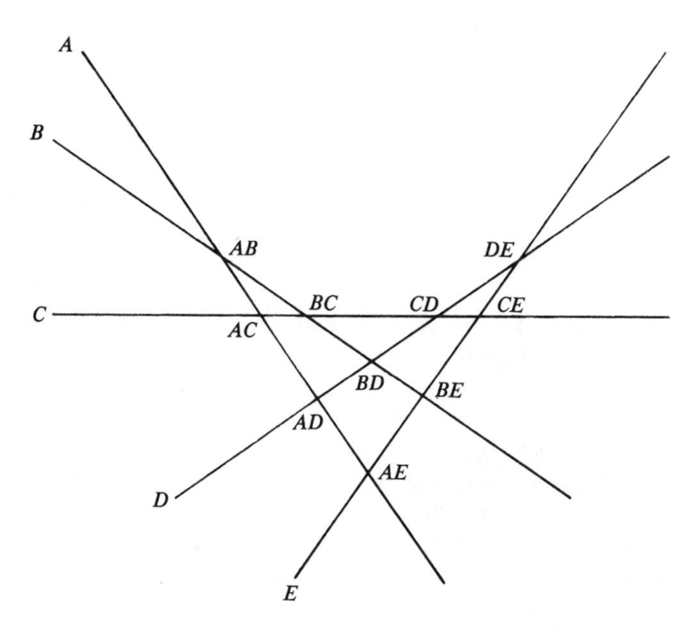

FIG. 1.19. Golay's five lines.

Label the first ten row positions of the matrix with the intersection labels from Figure 1.19. Then, for the column under Y_2, place a "0" in the positions corresponding to intersections of line A and a "1" elsewhere. Do the same for the column under Y_3 using line B, etc. The result is shown in Figure 1.20.

This completes the construction of the first ten row positions of columns Y_2 through Y_6 of Figure 1.17.

The column Y_7 is formed by associating a "0" with the five

		Y_2	Y_3	Y_4	Y_5	Y_6
1	AB	0	0	1	1	1
2	AC	0	1	0	1	1
3	AD	0	1	1	0	1
4	AE	0	1	1	1	0
5	BC	1	0	0	1	1
6	BD	1	0	1	0	1
7	BE	1	0	1	1	0
8	CD	1	1	0	0	1
9	CE	1	1	0	1	0
10	DE	1	1	1	0	0

FIG. 1.20

intersections AB, BE, ED, DC and CA, which are neighboring symbols in the 5-cycle

$$(ABEDC)$$

and 1's in the other five row positions. From the set of $4!-1$ such 5-cycles beginning with A, other than $ABEDC$ itself, pick the eleven that differ from $ABEDC$ by an even number of interchanges of elements. In this set of twelve, including $ABEDC$, there are six pairs which differ only in their order. One such pair is

$$ABEDC \quad \text{and} \quad ACDEB.$$

One representative of each pair is matched with one of the six Y_i's, $i = 7, 8, \ldots, 12$ as follows:

$$\begin{array}{ccc} Y_7 & Y_8 & Y_9 \\ (ABEDC) & (ABCED) & (ABDCE) \end{array}$$

$$\begin{array}{ccc} Y_{10} & Y_{11} & Y_{12} \\ (ACEBD) & (ACBDE) & (ADCBE) \end{array}$$

The columns associated with the eleven Y_i's, $i = 2, 3, \ldots, 12$, are displayed in Figure 1.21.

	Y_2	Y_3	Y_4	Y_5	Y_6	Y_7	Y_8	Y_9	Y_{10}	Y_{11}	Y_{12}
AB	0	0	1	1	1	0	0	0	1	1	1
AC	0	1	0	1	1	0	1	1	0	0	1
AD	0	1	1	0	1	1	0	1	0	1	0
AE	0	1	1	1	0	1	1	0	1	0	0
BC	1	0	0	1	1	1	0	1	1	0	0
BD	1	0	1	0	1	1	1	0	0	0	1
BE	1	0	1	1	0	0	1	1	0	1	0
CD	1	1	0	0	1	0	1	0	1	1	0
CE	1	1	0	1	0	1	0	0	0	1	1
DE	1	1	1	0	0	0	0	1	1	0	1

FIG. 1.21

The first column, Y_1, is chosen to be all 1's except for the last (the eleventh) element. Each of the columns $Y_2, \ldots, Y_{12}$ then receives an additional 1 in the eleventh place. Thus Y_1 differs from $Y_2, \ldots, Y_6$ in the necessary five places, and columns $Y_7, \ldots, Y_{12}$ each have the minimum of six 1's [**44**, p. 2]. This completes the construction of Golay's 11 by 12 check matrix given in Figure 1.17.

The calculation required to show that the various vector sums of columns of Golay's matrix satisfy the criteria of Table 1.18 is tedious but can be somewhat simplified by referring to the five lines in terms of which the columns were constructed. For example, inspection of Figure 1.19 shows that any two lines intersect the other three in six places. This means that the sum of any two of columns $Y_2, \ldots, Y_6$ in Golay's matrix has six 1's, exceeding the necessary number by one. Further checks are left to the reader.

Any linear code can also be viewed in terms of a *generating matrix* whose rows form a basis for the set of its codewords. For instance, the construction of the (23, 12) Golay code is much easier from this point of view. Any set $Y_1 \cdots Y_{12}$ of message digits is the linear combination, with coefficients 0 or 1 from the field GF(2), of the twelve row vectors

$$
\begin{array}{c}
100000000000 \\
010000000000 \\
\cdot \\
\cdot \\
\cdot \\
000000000001
\end{array}
\tag{1.22}
$$

Thus any codeword is a vector sum modulo 2 of the codewords formed from the above twelve sets of message digits. The first basis element has already been calculated as:

$$10000000000011111111110.$$

The second is

$$01000000000000001111111.$$

In general, the ith basis element is the row vector consisting of the ith vector in Table 1.22 followed by the ith column vector in Table 1.17. Thus, a 12 by 23 generating matrix for the (23, 12) Golay code is as shown in Figure 1.23.

$$
\begin{bmatrix}
1&0&0&0&0&0&0&0&0&0&0&0&1&1&1&1&1&1&1&1&1&1&0 \\
0&1&0&0&0&0&0&0&0&0&0&0&0&0&0&0&1&1&1&1&1&1&1 \\
0&0&1&0&0&0&0&0&0&0&0&0&0&1&1&1&0&0&0&1&1&1&1 \\
0&0&0&1&0&0&0&0&0&0&0&0&1&0&1&1&0&1&1&0&0&1&1 \\
0&0&0&0&1&0&0&0&0&0&0&0&1&1&0&1&1&0&1&0&1&0&1 \\
0&0&0&0&0&1&0&0&0&0&0&0&1&1&1&0&1&1&0&1&0&0&1 \\
0&0&0&0&0&0&1&0&0&0&0&0&0&0&1&1&1&1&0&0&1&0&1 \\
0&0&0&0&0&0&0&1&0&0&0&0&0&1&0&1&0&1&1&1&0&0&1 \\
0&0&0&0&0&0&0&0&1&0&0&0&0&1&1&0&1&0&1&0&0&1&1 \\
0&0&0&0&0&0&0&0&0&1&0&0&1&0&0&1&1&0&0&1&0&1&1 \\
0&0&0&0&0&0&0&0&0&0&1&0&1&0&1&0&0&0&1&1&1&0&1 \\
0&0&0&0&0&0&0&0&0&0&0&1&1&1&0&0&0&1&0&0&1&1&1
\end{bmatrix}
$$

FIG. 1.23. A generating 12 by 23 matrix for the (23, 12) Golay code.

This is simply $[I_{12} \vdots [a_{ij}]^T]$, where I_n denotes the identity matrix of order n. Showing that the rows in Matrix 1.23 generate the (23, 12) Golay code is the same as showing that the vector sum modulo 2 of any number of its rows has at least seven 1's. Since the vector sum modulo 2 of any seven or more rows clearly satisfies this criterion, we need only consider sums of six or fewer rows. This reduces to the calculations listed in Table 1.18 if the first twelve columns of Matrix 1.23 are deleted and the word "columns" is replaced by the word "rows" in the listed criteria.

By the way the generating matrix is formed, it is clear that the matrix is not unique. However, it has been shown [**101**], [**31**], that any binary block code with 2^{12} codewords of length 23 and minimum distance seven is equivalent to the (23, 12) Golay code. In other words, there is essentially only one way to tile the 2^{23} vertices of the unit 23-cube with Hamming spheres of radius three.

The second matrix Golay gave generates an (11, 6) perfect two-error-correcting ternary code. Note that entries in codewords are no longer restricted to 0 or 1, but can now be 0, 1 or 2. Spheres of radius two in this geometry have

$$\sum_{i=0}^{2} (3-1)^i \binom{11}{i} = 243 = 3^5$$

vertices. Since there are 3^{11} vertices, Golay saw that such a code might exist, and, of course, would have to consist of $3^{11}/3^5 = 3^6$ codewords. Again, without explaining its origin or showing that the generated code was two-error-correcting, he gave the following matrix (Figure 1.24):

	Y_1	Y_2	Y_3	Y_4	Y_5	Y_6
X_1	1	1	1	2	2	0
X_2	1	1	2	1	0	2
X_3	1	2	1	0	1	2
X_4	1	2	0	1	2	1
X_5	1	0	2	2	1	1

FIG. 1.24. Golay's check matrix for the (11, 6) code.

The extended matrix is

$$\begin{bmatrix} 1 & 1 & 1 & 2 & 2 & 0 & 1 & 0 & 0 & 0 & 0 \\ 1 & 1 & 2 & 1 & 0 & 2 & 0 & 1 & 0 & 0 & 0 \\ 1 & 2 & 1 & 0 & 1 & 2 & 0 & 0 & 1 & 0 & 0 \\ 1 & 2 & 0 & 1 & 2 & 1 & 0 & 0 & 0 & 1 & 0 \\ 1 & 0 & 2 & 2 & 1 & 1 & 0 & 0 & 0 & 0 & 1 \end{bmatrix} \tag{1.25}$$

This time the code can be viewed as the kernel of a linear transformation from the vector space $[GF(3)]^{11}$, with 3^{11} elements, to $[GF(3)]^5$ which has 3^5 elements. Since the code is linear, showing that the minimum distance between any two codewords is five reduces to showing that each nonzero linear combination of four or fewer columns of the extended matrix over GF(3) is not zero. Again, this criterion may be reduced to showing that linear combinations of four, three, two and one columns, over GF(3), of Golay's check matrix (Figure 1.24) have at least one, two, three or four nonzero entries, respectively. The total number of calculations is then

$$\sum_{i=1}^{4} (3-1)^i \binom{6}{i} = 472.$$

These are left to the reader. An equivalent check appears on pp. 90–91.

From the relation

$$\begin{bmatrix} X_1 \\ \cdot \\ \cdot \\ \cdot \\ X_5 \end{bmatrix} + [a_{ij}] \begin{bmatrix} Y_1 \\ \cdot \\ \cdot \\ \cdot \\ Y_6 \end{bmatrix} \equiv 0 \pmod 3$$

we see that

$$\begin{bmatrix} -X_1 \\ \cdot \\ \cdot \\ \cdot \\ -X_5 \end{bmatrix} \equiv [a_{ij}] \begin{bmatrix} Y_1 \\ \cdot \\ \cdot \\ \cdot \\ Y_6 \end{bmatrix} \pmod 3.$$

For instance, the encoding of

100000

is

10000022222,

while that of

010000

is

01000022110.

Thus, a generating matrix for the (11, 6) Golay ternary code is shown in Figure 1.26.

$$\begin{bmatrix} 1 & 0 & 0 & 0 & 0 & 0 & 2 & 2 & 2 & 2 & 2 \\ 0 & 1 & 0 & 0 & 0 & 0 & 2 & 2 & 1 & 1 & 0 \\ 0 & 0 & 1 & 0 & 0 & 0 & 2 & 1 & 2 & 0 & 1 \\ 0 & 0 & 0 & 1 & 0 & 0 & 1 & 2 & 0 & 2 & 1 \\ 0 & 0 & 0 & 0 & 1 & 0 & 1 & 0 & 2 & 1 & 2 \\ 0 & 0 & 0 & 0 & 0 & 1 & 0 & 1 & 1 & 2 & 2 \end{bmatrix}$$

FIG. 1.26. A 6 by 11 generating matrix for the (11, 6) Golay ternary code.

This is simply $[I_6 \vdots [-a_{ij}]^T]$.

As in the case of the (23, 12) code, this matrix clearly is not unique. However, it can be shown that any ternary block code with 3^6 codewords of length eleven and minimum distance five is equivalent to the one generated by this matrix. The proof is not short [**85**, p. 649].

No other perfect codes are known. This assertion must, however, be qualified by the recognition that there are non-linear perfect codes not equivalent to the Hamming codes but having the same parameters (word length, number of codewords, minimum distance). Finding all such codes is still an open problem. However, besides the trivial perfect codes—those consisting of just one codeword, or of all possible n-blocks of s symbols, or a binary repetition code of odd length—any perfect linear code over GF(p^m) must have the same parameters as one of the Hamming or Golay codes [**85**, p. 180]. Some nonexistence theorems are known for perfect codes of length n using s symbols where s is not a power of a prime. For example, even though $1 + (6 - 1)7 = 6^2$ happens to divide 6^7, there is no perfect single-error-correcting code of length seven using six symbols [**47**, p. 199]. The proof depends upon the nonexistence of orthogonal Latin squares of order six. For further information on perfect codes, see [**122**], [**85**, pp. 179–186].

Our discussion of the (23, 12) and (11, 6) Golay codes is not the approach taken in texts on coding theory, for there are much more elegant ways of generating these codes. Our purpose is, however, to develop the historical perspective. One problem of using Golay's work directly is showing that the proposed generating matrices do generate the codes, that is, that the minimum distances between the codewords generated by the matrices are indeed seven and five, respectively. Such a matrix for the (23, 12) Golay code, which can be shown to generate vectors at minimum distance more easily than "by inspection," will be given in Chapter 2. The reason for this is that Leech wanted one for his work in sphere packing [**76**].

Section 6. The Priority Controversy

Before we leave the history of the origins of coding theory and, in particular, the Hamming and Golay codes, we turn our attention briefly to questions of priority and nomenclature of the early codes. Although Kenneth O. May has suggested that "...the historian may be interested in studying priority controversies but should analyse historically significant events rather than award status," [**88**], we make this disgression because of some obscurity and controversy surrounding these "significant events."

Who is the discoverer of error-correcting codes—Golay, who published first, or Hamming, who was referenced by Shannon in 1948, a year earlier than Golay's paper of 1949? And furthermore, what should the various generalizations of the (7, 4) Hamming code be called? Berlekamp has considered both of these questions in some detail [**13,** pp. 2, 3, 7]. However, his remarks stimulated a response from Golay, who felt that Berlekamp had treated him unfairly [**43**].

As mentioned earlier, Hamming knew of the perfect binary codes prior to Shannon's publication in 1948, although neither Shannon nor Hamming could recall exactly when the codes were discovered [**56**], [**109**]. Shannon has suggested that it could have been as much as six months before his July, 1948, paper was published [**109**]. While Hamming's official publication in the *B.S.T.J.* was being held up for patent reasons, Golay read Shannon's work, which included the (7, 4) Hamming code, generalized it to the prime case, and published his work. He had, in fact, communicated with Shannon just prior to submitting his note for publication in 1949 [**39**]. Apparently Shannon had given a lecture on communication theory which Golay attended on January 19, 1949, after which they had discussed perfect binary coding. The next day, January 20, Golay wrote to Shannon,

> Permit me to thank you again for your interesting and stimulating talk of yesterday, and to congratulate you for your clearness of presentation of an arduous subject.
>
> Referring to our short conversation about binary cases which are codable without entropy increase, if we except the trivial cases of up to n errors in a package of $2n+1$ bits, and of one or no error in a package of 2^n-1 bit[s], I know of only one codable case, namely that of 23 bits, up to 3 of which can be in error.... [**39**]

Golay then gave the two coding matrices which were to appear in his 1949 paper, the (23, 12) and (11, 6) codes. He also showed why there could not be a (90, 78) systematic binary code and then asked Shannon if some of the digital electronic computers might be used in a search for other (nontrivial) perfect binary codes.

Shannon showed Golay's letter to Hamming. This explains Hamming's reference to Golay in both of the April 12, 1949, drafts of his paper [**52**], [**53**], [**109**].

As has been noted, Golay's 1949 paper was submitted on February 23, 1949, a little over a month after his letter to Shannon, and was published in June, two months after Hamming's first reference to Golay [**40**].

Who, then, gets the credit for first discovery? If the printed word or public announcement is the criterion, then Hamming should get credit for only the (7, 4) code and Golay for all the other binary perfect single-error-correcting codes as well as those in the nonbinary case—those Golay published as well as the ones Cocke discovered. Some authors have distributed credit in exactly this fashion, calling all the other single-error-correcting codes, other than the (7, 4), *H*-Golay codes [**7**, p. 247], [**8**, p. I-8]. In fact, the authors in the references just cited go on to say that:

> These codes are usually called Hamming codes, but Hamming invented only the one for $n = 7$, $q = 2$ [(7, 4) code]. . . . [**7**, p. 247], [**8**, p. I-8].

However, it has been shown that Hamming knew all the binary codes prior to Shannon's publication, and had circulated them in an interdepartmental memorandum several months prior to the submission date of Golay's one-page paper.

There is some justification for calling the nonbinary codes "H-Golay." Their discovery was inspired by Hamming's example, but Golay did the work. This name would, of course, have to be extended at least to those prime-power codes which Golay did discover, if not all of them (attaching the name H-G-Cocke to those Golay did not discover would be confusing indeed).

Unfortunately, this raises the specter of Hamming codes, H-Golay or H-G codes and Golay codes, all being distinct. Furthermore, the first two mentioned are very closely related. The historical solution, with the noted exceptions, has been to use "Hamming codes" to refer to all the single-error-correcting Hamming-type codes, whether binary or not, and to rserve the name "Golay" for the two perfect multi-error correcting codes which Golay also discovered.

Though perhaps this nomenclature is not the best solution, it was not meant to slight Golay. For even though Berlekamp concludes, "It is quite clear from the published literature that the subject of error-correcting codes was founded by Hamming (not Golay)," he goes on to describe, as we mentioned earlier, Golay's paper as the "best single published page." [**13**, p. 3, 4]

Our purpose is not to settle this undesirable dispute, but to recreate the moods and circumstances that surrounded the earliest years of the invention of coding theory. Having established these origins, we are now prepared to pursue the impact of coding theory on another branch of mathematics, namely, on the problem of how densely can congruent spheres be packed in Euclidean n-space, E^n. In particular, Golay's (23, 12) code plays the role of the initial linkup, overshadowing at first the contribution of any other code. But this is not all. Golay's code, through sphere packing, played yet a second role, that of stimulating the discovery in 1968 of three new simple groups of very large order. That link, however, shall have to wait until Chapter 3.

CHAPTER 2

FROM CODING TO SPHERE PACKING

Section 1. An Introduction to Sphere Packing

In Euclidean n-space, E^n, how may disjoint, open, congruent n-spheres be located to maximize the fraction of the volume of E^n that the n-spheres cover? That is the sphere packing problem, which goes back to a book review that Gauss wrote in 1831, in which he pointed out that a problem concerning the minimal nonzero value assumed by a positive definite quadratic form in n variables, first considered by Lagrange in 1773, could be translated into a problem on packing spheres (cf. C. A. Rogers [**105**, pp. 1, 106]).

Though there might seem to be no connection between coding theory and sphere packing, John Leech in [**72**] and [**74**] used results in coding theory to obtain a packing of E^{24} that was much denser than any previously known. The present chapter will describe his packing and discuss how he happened to bridge the gap, so to speak, from codes to sphere packings.

The (23, 12) Golay code can be extended to a (24, 12) code by adding a 0 or 1 to each codeword so that all codewords have even weight. This (24, 12) Golay code thus has a minimum codeword distance of eight. Leech's packing of E^{24} employs* the (24, 12) Golay code and has a density of

---

*The word "employs," though convenient, is misleading. Leech was only aware of the (23, 12) Golay code. Knowing that there is one led him to search for a generating matrix for a code in twenty-four dimensions which could be readily checked. Later it was shown that all such codes in twenty-three (and twenty-four) dimensions are equivalent [**101**].

0.001929, nearly seventy-nine percent of the upper bound 0.002455 for density of sphere packings in E^{24}, given by Rogers [**74**]. In fact, this packing's density is far closer to its corresponding upper bound than known packings in E^n for nearby dimensions are to their respective upper bounds. (See Table A1.1 of Appendix 1.)

The centers of the spheres of this packing form a lattice (which we shall define later), called the "Leech lattice," which in turn led Conway to his discovery of new simple groups, described in Chapter 3.

In the papers [**72**], [**74**] that Leech devoted to sphere packing, a mere three of the forty-three pages are spent describing this lattice packing. Of such brevity, Leech says [**77**]:

> As this [packing] was almost a 'throw-away,' its importance was not realized until Conway started exhibiting fascinating groups associated with it. (In comparison, the whole of 1967 part 1 I now regard as insignificant.)*

In fact, Leech himself suspected that the group of symmetries of his lattice would contain some large simple groups. But this is getting a little ahead of our story. First, we present some definitions and some examples of sphere packings.

Let S be an open n-dimensional sphere of radius r and (n-dimensional) volume $V_n r^n$, centered at the origin, where V_n is a constant dependent only on n. Let $\mathbf{a}_1, \mathbf{a}_2, \mathbf{a}_3, \ldots,$ be an infinite sequence of points (vectors) in E^n. The set of translates of S, $\mathbf{a}_1 + S$, $\mathbf{a}_2 + S$, $\ldots$, form a *packing* when they are pairwise disjoint. In case the centers, $\mathbf{a}_1, \mathbf{a}_2, \ldots,$

---

*Professor Leech wishes it to be known that his letters were written without anticipating that extracts from them would be published and that they do not necessarily represent his style and standards for publication; and, furthermore, that he retains the copyright in and does not authorize the further reproduction of such extracts.

form a group under vector addition which has dimension n, the packing is called a *lattice packing*. In this case it can be shown that there are n linearly independent vectors, $\mathbf{v}_1, \ldots, \mathbf{v}_n$ such that the centers $\mathbf{a}_1, \mathbf{a}_2, \ldots,$ are integral combinations of the $\mathbf{v}_i$.

The *density* of packing is the fraction of E^n covered by the spheres. To be more precise (and following [**105**]), let K denote a "half-open, half-closed" n-cube with its edges parallel to the coordinate axes. Thus, K is the set of all vectors $\mathbf{v}$ in E^n, $\mathbf{v} = (x_1, x_2, \ldots, x_n)$, such that

$$\begin{aligned} c_1 - \tfrac{1}{2}s &\leq x_1 < c_1 + \tfrac{1}{2}s, \\ c_2 - \tfrac{1}{2}s &\leq x_2 < c_2 + \tfrac{1}{2}s, \\ &\vdots \\ c_n - \tfrac{1}{2}s &\leq x_n < c_n + \tfrac{1}{2}s, \end{aligned}$$

where $(c_1, c_2, \ldots, c_n)$ is the center of the cube with edge length s. Let $P = \{\mathbf{a}_1 + S, \mathbf{a}_2 + S, \ldots\}$. Then write

$$\rho_+(P, K) = \frac{1}{s^n} \sum_{(\mathbf{a}_i + S) \cap K \neq \emptyset} V_n r^n$$

$$\rho_-(P, K) = \frac{1}{s^n} \sum_{(\mathbf{a}_i + S) \subset K} V_n r^n.$$

Thus, $\rho_+(P, K)$ is the ratio of the total volume of those spheres in P that have a point in common with K to the volume of K, while $\rho_-(P, K)$ is the ratio of the total volume of those spheres contained in K to the volume of K. The *upper density* of P is defined as

$$\rho_+(P) = \lim_{s \to \infty} \sup_K \rho_+(P, K)$$

while the *lower density* of P is

$$\rho_-(P) = \lim_{s \to \infty} \inf_K \rho_-(P, K).$$

Clearly $\rho_-(P) \leq \rho_+(P)$. When $\rho_-(P) = \rho_+(P)$, their common value is called the *density* of the packing and is denoted ρ. In the case of lattice packings it is easy to see that the density ρ exists and it is usually a routine exercise to compute it. It is the volume of a sphere divided by the amount of space nearer to its center than any other. A related quantity, the *center density*, δ, is the density of the packing divided by V_n. The number δ is proportional to the density, ρ, and is the term usually used in the literature. However, we shall use ρ for all our calculations.

For example, consider E^1 and the lattice packing of spheres of radius r with centers of the form:

$$2zr, \quad z \in Z,$$

where Z is the set of integers. This tiling of the line by abutting open intervals of length $2r$ is pictured in Figure 2.1.

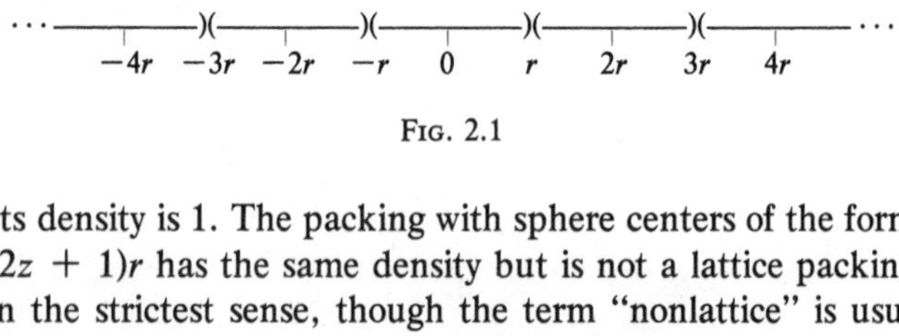

$-4r \quad -3r \quad -2r \quad -r \quad 0 \quad r \quad 2r \quad 3r \quad 4r$

FIG. 2.1

Its density is 1. The packing with sphere centers of the form $(2z + 1)r$ has the same density but is not a lattice packing in the strictest sense, though the term "nonlattice" is usually restricted to packings that do not satisfy the requirements for a lattice even with a sphere center at the origin. Both are clearly densest possible packings.

The plane, E^2, is more interesting. In this case, the "volume of a sphere of radius r," V_2r^2, is just its area, πr^2, and so $V_2 = \pi$. For convenience, let $r = 1$. The lattice basis $\mathbf{v}_1 = (2, 0)$ and $\mathbf{v}_2 = (0, 2)$ provides the "square" lattice packing depicted in Figure 2.2.

By elementary geometry, its density is $\pi/4 = 0.7853$.

A denser packing is obtained by using the lattice generated by $\mathbf{v}_1 = (2, 0)$ and $\mathbf{v}_2 = (1, \sqrt{3})$. This is shown in Figure 2.3.

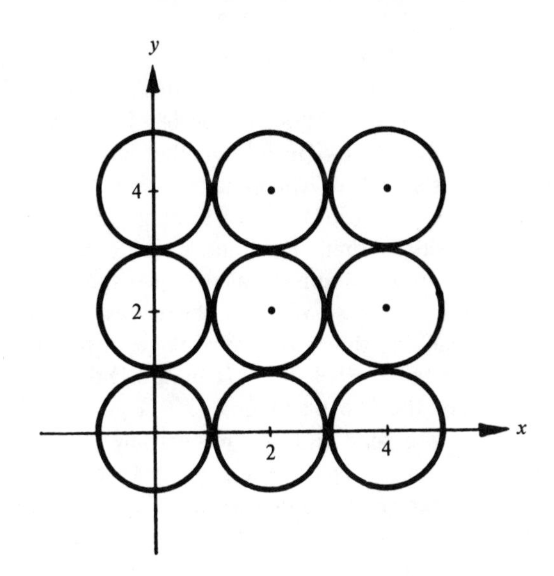

FIG. 2.2. "Square" lattice packing in E^2.

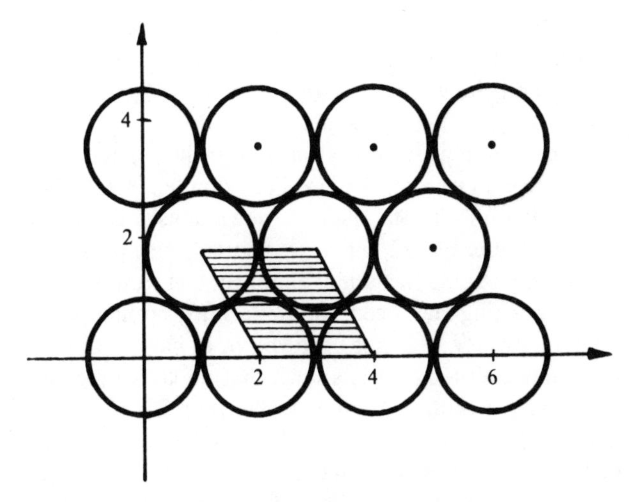

FIG. 2.3. Densest packing in E^2.

Observe that the shaded parallelogram of height $\sqrt{3}$ and width 2 contains the equivalent of one disk. Thus the density of this packing is $\pi/(2\sqrt{3}) = 0.9068$. Incidentally, no packing of congruent disks in E^2, whether lattice or not, is denser than this so-called "triangular packing."

Closely related to the problem of finding packings in E^n with large density is that of finding packings in E^n with large *contact number*. Assume that the open spheres are replaced by their closures. What is the largest number of spheres in a packing that any given sphere in the packing may touch? For the example in E^1 (Figure 2.1) the answer is clearly two. For the two packings in E^2 (Figures 2.2, 2.3) these numbers are four and six, respectively. In a lattice packing, this number is the same for any sphere in the packing. However, for a nonlattice packing, this number may vary from sphere to sphere. To see this, simply remove a finite number of spheres from a given lattice packing. Doing so changes the contact number of a finite number of spheres, but does not change the density. In case this number is different for more than a finite number of spheres, we will speak of both the *maximum contact number* of the packing and the *average contact number.* For lattice packings, these quantities coincide.

Lattice packings do not always have the highest contact numbers for a given dimension. For example, in E^9, the lattice packing Λ_9 with the highest known density 0.1457 has a contact number of 272, whereas the nonlattice packing $P9$a with density 0.1288 has a maximum contact number of 306. Neither packing has been shown to have optimal density. Furthermore, $P9$a has not been shown to have optimal contact number although it is known that no lattice packing in E^9 can have a contact number higher than 272 [**83**, p. 725], [**123**]. It should be noted that the average contact number of $P9$a is 235.6, less than that of Λ_9. However, in E^{10} there is a less dense nonlattice packing with greater average contact number than the densest known lattice packing in that dimension (see Appendix 1).

We have already mentioned Rogers' upper bound for the density of a packing in E^n and will now take a closer look at it. The convex hull of $n + 1$ points, each pair the same fixed distance apart, is called a regular n-simplex. For $n = 1, 2, 3$, we get the following well-known figures:

$n = 1$ line segment

$n = 2$ equilateral triangle

$n = 3$ regular tetrahedron.

Rogers showed in 1958 that the density of a packing in E^n cannot exceed the fraction of a regular n-simplex of side length two which is interior to the $n + 1$ unit spheres whose centers are the vertices of the simplex [**104**]. Figure 2.4 shows the case for $n = 2$.

FIG. 2.4

From Figures 2.1 and 2.3, it is easily seen that this bound is attained in both the packing given in E^1 and the second packing given in E^2. For dimension n, this bound of Rogers is

$$\frac{(n+1)^{1/2}(n!)^2\pi^{n/2}}{2^{3n/2}\Gamma\left(\frac{n}{2}+1\right)} f_n(n).$$

In this expansion $\Gamma(x)$ is the Gamma function, $\int_0^\infty t^{x-1}e^{-t}dt$. To obtain $f_n(n)$, first define $F_n(\alpha)$ recursively by

$$F_{n+1}(\alpha) = \frac{2}{\pi}\int_{\frac{\operatorname{arcsec}(n)}{2}}^{\alpha} F_{n-1}(\beta)d\theta,$$

$\sec 2\beta = \sec(2\theta) - 2$ and $F_1(\alpha) = F_0(\alpha) = 1$. Then $f_n(\sec 2\alpha) = F_n(\alpha)$ [**30**, p. 53], [**72**, p. 675]. Calculations of $f_n(n)$, for $n \leq 8$, which involve numerical integration, are given in [**30**, pp. 67, 68] where they are credited to Leech.

The densest known packing in E^3 can be formed by taking the "square" packing in E^2, replacing the disks by spheres, and then placing these layers, one upon another, so that each sphere in a layer touches four spheres in the layer above and four in the layer below. The density of this packing, which will be calculated shortly, is less than the bound given by Rogers. No one has found a denser packing in E^3 nor shown that this one is optimal [**5**, p. 100].

This lattice packing is most easily generated by the vectors $\mathbf{v}_1 = (1, 1, 0)$, $\mathbf{v}_2 = (1, 0, 1)$ and $\mathbf{v}_3 = (0, 1, 1)$. The corresponding spheres are of radius $1/\sqrt{2}$. Figure 2.5 shows a cross section of the packing in the xy plane.

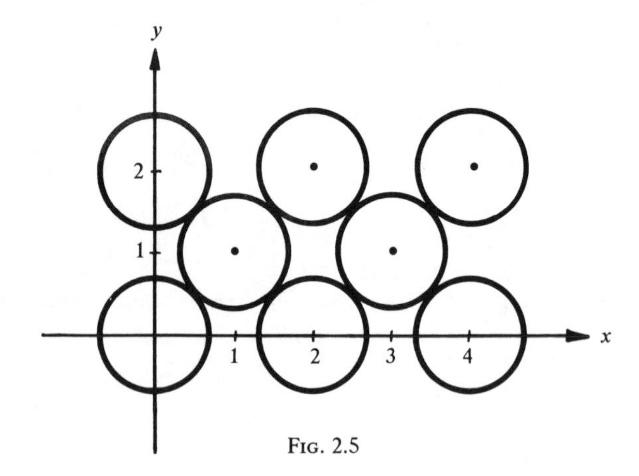

FIG. 2.5

This is simply the "square" lattice packing of E^2 (with spheres of reduced radii) rotated 45° (see Figure 2.2). The same cross section appears in the other coordinate planes.

One way to calculate the density is to note that the unit cube, whose vertices consist of 0's and 1's, contains four sphere octants, i.e., half a sphere. By repeated reflections in its face planes, this cube generates the entire packing. Thus each sphere has associated with it the volume of two unit cubes and the density of the packing is

$$\frac{4\pi(1/\sqrt{2})^3/3}{2} = 0.7404.$$

Incidentally, Rogers' upper bound for the density of any packing in E^3 is 0.7796.

Figure 2.6 shows a cross section of this packing with sphere centers in the plane $x + y + z = 2$.

As a planar section, this is equivalent to the densest packing in E^2 (see Figure 2.3). Thus, the densest known packing in E^3 can be thought of as being made up of layers of the densest packing ("triangular") in E^2. However, the orientation given eliminates having to use sphere centers with non-integer coordinates as would be the case if the "triangular" packing layers were placed parallel to any of the coordinate planes.

If one were to view the sphere centers from "infinity" along a line perpendicular to the $x'z'$ plane of Figure 2.6, one would see three distinct layers, which are shown in Figure 2.7.

The sphere centers denoted by "1" are those in the plane $x + y + z = 2$, centers denoted by "2" are those in the plane $x + y + z = 4$ and centers denoted by "3" are those in the plane $x + y + z = 6$. The next layer of centers would line up with the first layer and so on. A nonlattice packing of the same density is formed by alternating the first and second of these layers infinitely often, with period two.

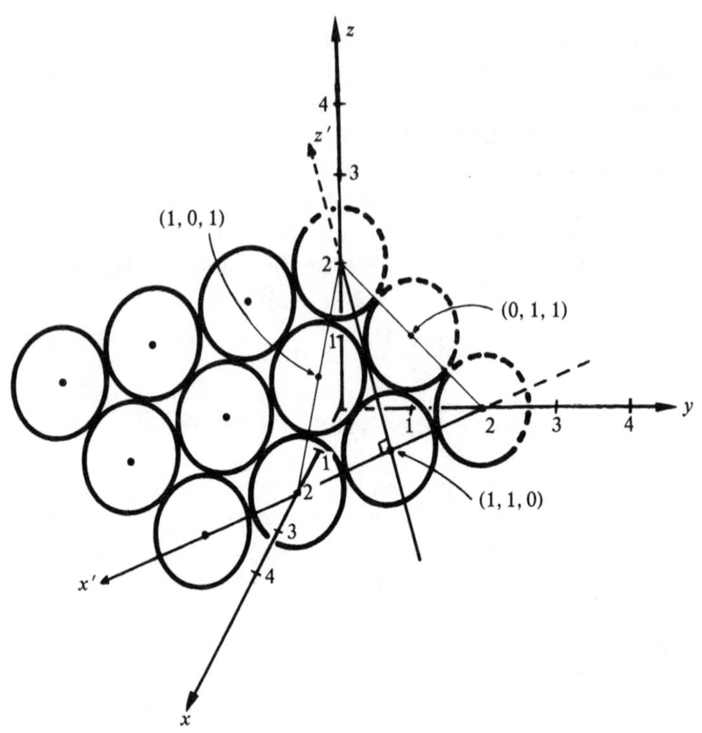

FIG. 2.6. A skew layer of the densest lattice packing in E^3 in the plane $x + y + z = 2$.

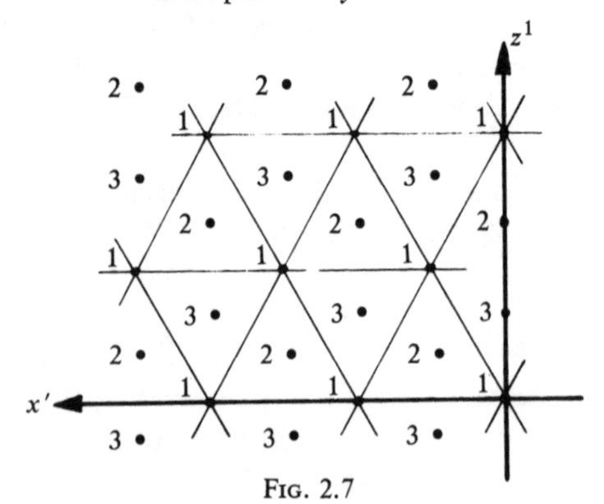

FIG. 2.7

The contact number of either of the two packings is twelve, since any sphere touches six others in its own skew layer, three in the layer below and three in the layer above. Alternatively, for the lattice packing, any sphere touches four others in its own square layer, four in the layer below and four in the layer above.

Let us calculate the density of this same packing by a method that will be used later in the study of packings in E^8 and E^{24}. Recall that the lattice is defined to be all integer combinations of the vectors: $\mathbf{v}_1 = (1, 1, 0)$, $\mathbf{v}_2 = (1, 0, 1)$ and $\mathbf{v}_3 = (0, 1, 1)$. It is an easy exercise to show that the lattice consists of exactly those points in E^3 with integer coordinates whose sum is even. This is, of course, half of the points with integer coordinates. Thus we get, on the average, one-half sphere center per unit volume and the density calculation yields

$$\frac{1}{2} \cdot \frac{4\pi(1/\sqrt{2})^3}{3} = 0.7404$$

which is the same as before.

With these details in order, we are ready to examine the transition from coding to sphere packing. Moreover, we shall examine how this transition happened to reveal the circuitous ways that mathematical ideas develop and leap from one field to another. These jumps, so important in the growth of mathematics, are sometimes impossible to reconstruct from the published literature, whose austere, disembodied style focuses on elegantly presented results rather than on their genesis.

Section 2. The Leech Connection

Leech's first paper on sphere packing in higher dimensions appeared in 1964 [**72**], where he presented new packings in several dimensions larger than eight, including a packing in E^{24}, utilizing the existence of the (23, 12) Golay code, and a comparison of the densities and contact numbers of these packings with Rogers' packing bound and Coxeter's proposed contact number bound. There he also credited the work of H. F. Blichfeldt in 1929 and 1935 for establishing the densest lattice packings in dimensions $n \le 8$ [**72**, p. 657], [**18**], [**19**]. However, his own interest in sphere packings and other geometric problems was kindled during his student days at Cambridge, England, where he received his B.A. in 1950 [**75**]. Of this period he writes:

> My strongest personal stimulus was Professor H. S. M. Coxeter, from his books, papers and [later] correspondence.... My acquaintance with Coxeter and his work was via his book *Regular Polytopes* (1948) [**29**].... My acquaintance with Golay's codes is also of this period, probably via the reference in Shannon and Weaver's [*The*] *Mathematical Theory of Communication* (1949) [the latter book is a reprint of Shannon's 1948 paper—with corrections and additional references (notably Golay's 1949 paper) —and an expanded version of a paper by Warren Weaver which appeared in *Scientific American*, July, 1949 [**110**, preface, p. 48], [**124**]]... I do not now recall my first acquaintance with either Shannon and Weaver's book or Golay's codes. I am an omnivorous browser in libraries and book shops, and either might have been at any time. [**75**], [**76**], [**78**].

Leech knew of the Golay codes perhaps as early as 1949, certainly before 1951. Now we will examine how he exploited the existence of the (23, 12) code to construct the densest known sphere packing of E^{24}. As we describe Leech's contribution, we will endeavor to recreate his thought processes.

One way to generate a sphere packing of E^n is to specify that the coordinates of n linearly independent vectors be the rows of an n by n matrix and then take, as the sphere centers of the packing, all possible linear combinations, with integer coefficients, of these rows. From now on, whenever "coordinates" are mentioned for a packing, they are those of the sphere centers. The congruent spheres in a packing will usually not have radius one. Furthermore, to simplify all calculations, the entries of the matrix will be integers. Of course, with this latter requirement, we lose generality. In particular, the densest packing in E^2 does not meet this requirement. However, good (dense) packings are possible. In fact, we can make them as close as we wish to the best since the rational numbers are dense in the reals.

If the minimum Euclidean distance between any pair of such vectors is say, t, the matrix *generates* a packing of spheres whose radii are $t/2$. For example, consider the matrix

$$\begin{bmatrix} 1 & 0 \\ 0 & 1 \end{bmatrix}.$$

Since the minimum distance between generated vectors is one, this matrix defines a packing of spheres, in E^2, of radii one-half, which is just a scaled-down version of the "square" packing in Figure 2.2. For larger n, it may be difficult to find the minimum distance between the vectors generated by the matrix.

One orientation of the densest lattice packing in E^4 has as sphere centers the set of all quadruples of either all even or all odd integers and is generated by the matrix

$$\begin{bmatrix} 1 & 1 & 1 & 1 \\ 2 & 0 & 0 & 0 \\ 0 & 2 & 0 & 0 \\ 0 & 0 & 2 & 0 \end{bmatrix}.$$

It can be generalized directly to E^8. Although this packing, which we shall call L_8, is not the best lattice packing in E^8, it illustrates both how to find the contact number and the density. Since this is a lattice packing, the contact number is the same for all spheres and thus is the number of lattice points at the minimum distance from the origin. (The only spheres that touch the sphere at the origin are those with centers at these points.) One point closest to the origin is

$$(2, 0, 0, 0, 0, 0, 0, 0).$$

Other points closest to the origin consist of seven 0's and one 2 or -2 at any of the eight positions. Such a point has *shape* $(0^7, \pm 2)$. Thus the contact number is sixteen.

The center density is fairly easy to calculate. If all coordinate entries were written in binary form, then the lattice, by definition, would contain only those coordinates whose ones digits were either all 0's or all 1's. In the case of E^8 this means that only two out of every 2^8 points with integer coordinates are acceptable. Thus the center density is $1/(2^7)$.

Before we can calculate the density of this packing in E^8, we pause to show how to calculate the volume of a sphere in E^n by elementary calculus.

Consider the n-sphere of radius r and center at the origin defined by

$$x_1{}^2 + x_2{}^2 + \cdots + x_n{}^2 \leq r^2.$$

For $n = 1$, its volume is $2r$. For $n = 2$, its volume is

$$\int_{-r}^{r} 2\sqrt{r^2 - x_1{}^2}\, dx_1 = \pi r^2.$$

This is obtained by integrating from $-r$ to r the one-dimensional volume of 1-spheres of radius $\sqrt{r^2 - x_1{}^2}$. This radius is that of the largest 1-sphere whose center is at the point x_1 on the x_1 axis and which lies in the 2-sphere (see Figure 2.8).

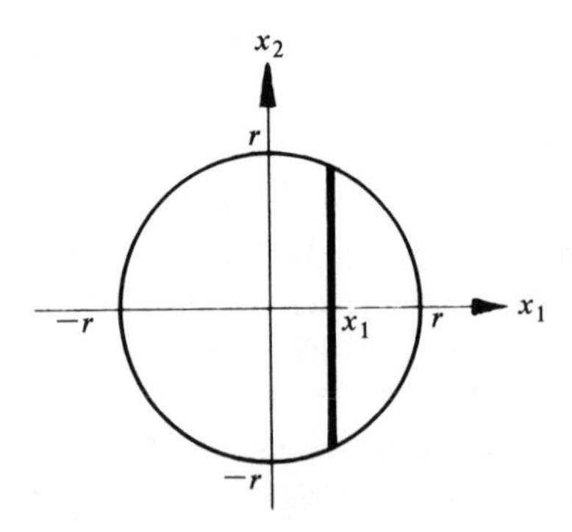

FIG. 2.8

For $n = 3$, the volume is

$$\int_{-r}^{r} \pi(r^2 - x_1^2)dx_1 = \frac{4}{3}\pi r^3.$$

Again, this is obtained by integrating from $-r$ to r the two-dimensional volume of 2-spheres of radius $\sqrt{r^2 - x_1^2}$ by the cross-section method for finding volume. Similarly, for $n = 4$ the volume is

$$\int_{-r}^{r} \frac{4}{3}\pi(r^2 - x_1^2)^{3/2}dx_1 = \frac{\pi^2 r^4}{2}.$$

If the constant factor in the volume formula of the n-dimensional sphere of radius r is denoted V_n, we obtain the general formula

$$\int_{-r}^{r} V_{n-1}(r^2 - x_1^2)^{\frac{n-1}{2}} dx_1 = V_n r^n$$

$$= \text{Volume of } n\text{-dimensional sphere of radius } r.$$

It can be shown by induction that

$$V_n = \frac{\pi^{n/2}}{\Gamma\left(\frac{n}{2} + 1\right)}$$

where $\Gamma(x)$ is the Gamma function. In particular, for $n = 8$ the 8-dimensional volume is

$$\frac{\pi^4 r^8}{4!}.$$

The spheres in the given packing in E^8 are of radius one; that is, half the distance of (2, 0, 0, 0, 0, 0, 0, 0) from the origin. Since the center density is $1/2^7$, the density of this packing is

$$\frac{\pi^4 1^8/4!}{2^7} = 0.03170,$$

which is much smaller than Rogers' bound, 0.2567.

Now delete some of the vertices and at the same time increase the radii of the remaining spheres. Specifically, keep only those vertices of the given lattice, L_8, whose coordinates sum to a multiple of four. This time the closest points to the origin are the $2^7 = 128$ points with shape $(\pm 1^8)$, points whose entries are either 1 or -1 and sum to a multiple of 4, together with the $4 \cdot \binom{8}{2} = 112$ points of shape $(0^6, \pm 2^2)$. Thus, the contact number equals $128 + 112 = 240$.

By imposing the additional requirement that the sum of the coordinates be a multiple of four, we have eliminated half of the centers of the original packing. However, the points with shapes $(\pm 1^8)$ and $(0^6, \pm 2^2)$ are now the closest centers to the origin. This means that the radius of each sphere can be expanded to $\sqrt{8}/2 = \sqrt{2}$. Each sphere in this new packing has $(\sqrt{2})^8/1^8 = 16$ times the volume of spheres in the lattice packing L_8. Since there are half as many per unit volume, the density is eight times that of the previous packing, or

$$\frac{\pi^4}{4! \cdot 2^4} = 0.2536.$$

This is the densest known packing, called Λ_8, of E^8 by congruent spheres. Its density is quite close to Rogers' bound, 0.2567. Leech noted that "this construction is less successful in producing dense lattice packings in E^{2m} for $m > 3$, even with this restriction on the sum of the coordinates." [**72**, p. 658]

Leech gave a second construction of the densest lattice packings in E^4 and E^8, which we describe because it leads to his packing in E^{24}. He considered matrices defined recursively by the equations

$$A_1 = 0, A_{2n} = \begin{bmatrix} A_n & A_n \\ A_n & \overline{A_n} \end{bmatrix}$$

where $\overline{A_n}$ denotes the matrix with entries of the form $\overline{a_{ij}} = 1 - a_{ij}$, a_{ij} being the corresponding entry of A_n [**72**, p. 658]. For $n = 1$, 2 and 3 these matrices are therefore:

$$A_2 = \begin{bmatrix} 0 & 0 \\ 0 & 1 \end{bmatrix}$$

$$A_4 = \begin{bmatrix} 0 & 0 & 0 & 0 \\ 0 & 1 & 0 & 1 \\ 0 & 0 & 1 & 1 \\ 0 & 1 & 1 & 0 \end{bmatrix}$$

$$A_8 = \begin{bmatrix} 0 & 0 & 0 & 0 & 0 & 0 & 0 & 0 \\ 0 & 1 & 0 & 1 & 0 & 1 & 0 & 1 \\ 0 & 0 & 1 & 1 & 0 & 0 & 1 & 1 \\ 0 & 1 & 1 & 0 & 0 & 1 & 1 & 0 \\ 0 & 0 & 0 & 0 & 1 & 1 & 1 & 1 \\ 0 & 1 & 0 & 1 & 1 & 0 & 1 & 0 \\ 0 & 0 & 1 & 1 & 1 & 1 & 0 & 0 \\ 0 & 1 & 1 & 0 & 1 & 0 & 0 & 1 \end{bmatrix}.$$

Such matrices were derived from those studied by J. J. Sylvester and R. E. A. C. Paley [**72**, p. 658], [**118**], [**99**]. Paley's matrices have 1's and -1's where Leech's have 0's and 1's, respectively, and also have the property that

$$HH^T = nI.$$

This matrix equation defines a *Hadamard matrix of order n*, where H has only 1's and -1's as entries. (The order of a Hadamard matrix is necessarily 1, 2 or a multiple of 4. It is not known whether there is a Hadamard matrix of all such orders.)

To define the lattice packing, Λ_8, in E^8, Leech took all octuples whose coordinates are congruent modulo 2 to the coordinates of the rows in either A_8 or $\overline{A_8}$. Although the lattice of this packing differs in orientation from that previously given for Λ_8, there is a simple relation between the coordinates. Let $(x_1, \ldots, x_8)$ be a vector of the lattice previously given for Λ_8. A point in the reoriented lattice will then have coordinates $(y_1, \ldots, y_8)$ where, for $i = 1, 2, 3, 4$,

$$y_{2i} = \tfrac{1}{2}(x_{2i-1} - x_{2i})$$
$$y_{2i-1} = \tfrac{1}{2}(x_{2i-1} + x_{2i}).$$

The rows of A_8 and $\overline{A_8}$ have the property that they differ from each other by either exactly four or eight places. In general, the rows of A_n and $\overline{A_n}$ differ from each other in either exactly $n/2$ or n places. For dimension eight, this suggests a code with sixteen codewords and minimum Hamming distance four. Such a code is known as the first order Reed-Muller code of length eight [**85**, p. 373]. It is generated by all possible vector sums modulo 2 of the four vectors:

$$\begin{matrix} 0 & 0 & 0 & 0 & 1 & 1 & 1 & 1 \\ 0 & 0 & 1 & 1 & 0 & 0 & 1 & 1 \\ 0 & 1 & 0 & 1 & 0 & 1 & 0 & 1 \\ 1 & 1 & 1 & 1 & 1 & 1 & 1 & 1 \end{matrix}$$

A family of such codes was constructed by I. S. Reed and D. E. Muller in 1954 [**93**], [**102**].

In passing, it should be noted that the analogous packing in E^{16} is not the densest known. For example, the closest points to the origin are those with exactly one ± 2 and fifteen 0's. All points with odd coordinates are farther away from the origin. The contact number is thus 32 while the density is

$$\frac{2^5}{2^{12}} \cdot \frac{\pi^8 1^{16}}{8!} = 0.0001149.$$

Leech increased the density by retaining only those vertices whose coordinate sum is divisible by 4, the same device as that in going from L_8 to Λ_8 [**72**, p. 659]. This cuts the number of centers in half, increases the contact number to 4320 and, since the radius of each sphere is now $\sqrt{2}$, boosts the density to

$$\frac{1}{2^{12}} \cdot \frac{\pi^8 (\sqrt{2})^{16}}{8!} = 0.0470,$$

which is 128 times the old density. Leech was to use this restricting device for the same reason in his first packing in E^{24} [**72**, p. 671], [**82**].

Leech then introduced the notion of *k-parity* to generalize these constructions in E^8 and E^{16} to form good lattice packings in E^{2^n}. In doing so in his 1964 paper, he rediscovered all the Reed-Muller codes. But he was unaware of any connection with them until after his paper was received on June 26, 1963, by the publishers [**72**, p. 657].

Sometime between June 26 and November 2, 1963, Leech came upon W. W. Peterson's *Error-Correcting Codes*, published in 1961. It was one of the first two texts on coding theory and contained a description of the Reed-Muller codes on pages 73–77 [**100**, preface]. Leech was at Glasgow University at the time and discovered the book on a colleague's shelf in a temporarily shared office [**75**], [**76**].

At the end of the 1964 paper in a section titled "Notes added November 2, 1963," Leech acknowledges his discovery of Peterson's book and the anticipation of the results in the first few sections of his paper by the Reed-Muller codes [**72**, p. 681].

The important idea in this second construction of the densest known lattice packing in E^8 is that Hadamard-type matrices, A_8 and $\overline{A}_8$, generate a good lattice packing.

Leech's third construction of this same packing in E^8 is closely linked to the coding theory discussed in Chapter I. He noted that in E^7 each vertex of the unit 7-cube had seven neighbors at Euclidean (and Hamming) distance one from itself [**72**, p. 668]. Furthermore, this number, eight, of points in any such neighborhood divides the total number of vertices of the 7-cube. Said Leech, "Since this is an exact submultiple of the total number of vertices, we may ask whether or not we can find a subset of the vertices which are centres of a set of neighbourhoods which exactly exhaust the vertices of the cube." [**72**, p. 668] This was immediately followed by the affirmative answer: When the same column in both A_8 and $\overline{A}_8$ is deleted, the septuples consisting of the sixteen rows in the two 8 by 7 matrices so formed do the job. Furthermore, the column to be deleted is arbitrary. This is easy to see from the fact that the rows of A_8 and $\overline{A}_8$ differ from each other in either four or eight places. Once this set of sixteen septuples has been found, the deleted coordinate is restored by parity and a lattice is formed by congruence modulo 2. This gives the densest known lattice packing Λ_8, of E^8, as before.

Note that the sixteen septuples above must be equivalent to the (7, 4) Hamming code since they form a vector space over GF(2). The restoration of the deleted coordinate is the equivalent of adding an overall parity check digit to the (7, 4) Hamming code, thereby generating the extended (7, 4) Hamming code. This is not to be confounded with the (8, 4) Hamming code. This extended code detects any two errors as well as corrects any single error since the minimum codeword distance is four.

Although Leech generalized this process to any space E^{2m}, he found a slightly different approach more fruitful. As he put it:

> We consider next the possibility of neighbourhoods of vertices distant greater than 1 on the unit cube γ_n. The number of such vertices within distance r (measured along the edges) is
>
> $$1+n+\binom{n}{2}+\cdots+\binom{n}{r},$$
>
> and we have to find such sums which are powers of 2. There are trivial solutions with $r \geq n$ (useless for our purposes) and with $r = \frac{1}{2}(n-1)$, leading to the packing with all coordinates odd or all even, which is of low density for $n > 4$. [**72**, p. 669]

That brief, rich paragraph presents the same counting argument used by Golay, the Hamming distance function, and the necessary condition for the existence of the two-word perfect ϵ-error-correcting binary codes, $n = 2\epsilon + 1$.

Leech claimed, for $\epsilon = 2$, that the only possibility for an exact 2-cover is in E^{90} but that Golay's 1949 paper and a paper by Lowell J. Paige in 1957 showed that the cover could not exist [**72**, p. 669], [**40**], [**96**, p. 18]. (An *exact ϵ-cover* is a set of vertices of the unit cube in E^n such that Hamming spheres of radius ϵ centered on these vertices exactly fill out the vertices of the n-cube with no overlap, i.e., a perfect ϵ-error-correcting binary code with word length n.) Actually, Golay's paper shows only that a *linear* exact 2-cover could not exist in E^{90}. Paige simply states:

> For example, $N(2, 90, 2) = 2^{12}$ [recall that $1 + \binom{90}{1} + \binom{90}{2} = 2^{12}$] yet a simple analysis of vectors having three nonzero coordinates shows that no exact 2-covering can exist. [**96**, p. 18]

Leech noted that for $\epsilon = 3$, $1 + 23 + \binom{23}{2} + \binom{23}{3} = 2^{11}$ and, therefore, an exact 3-cover may exist for $n = 23$. Later in the paper Leech exhibited one such cover and used it to construct a new packing in E^{24}.

Section 3. The Origin of Leech's First Packing in E^{24}

How did Leech generate his first packing in E^{24}? He needed a systematic way of arranging the sphere centers, and his solution was to find a matrix of 0's and 1's whose twelve rows, and all vectors in the row space (i.e., all vector sums of rows modulo 2) differed from each other in at least eight places. The lattice was then defined to be the set of points in E^{24} whose coordinates are congruent, modulo 2, to the coordinates of points in the row space of the matrix and whose coordinates have sum divisible by four.

Finding such a matrix was not easy. However, given such a matrix, one could then delete any single column and have a 12 by 23 matrix with any two vectors in the row space at minimum distance seven. This would then be an exact 3-covering of the unit 23-cube. The original matrix is recovered by adding an overall parity check digit to each row. Golay essentially gave such a 12 by 23 matrix in generating his (23, 12) code (see Figure 1.17). This Leech knew. And the form he could have used appears in Figure 1.23. However, we have again a case where the exposition in the technical paper remains but most of the history is lost. Says Leech:

> I don't now recall at what stage I realized that the Golay code and everything else belonged in 24 dimensions rather than 23. Having done so, I did all the work with 24 and let 23 appear by the side. [**82**]

Leech also knew that Paige's paper [**96**], which concerned a matrix representation of the Mathieu group M_{23}, gave a matrix whose rows would generate an exact 3-covering of the unit 23-cube [**72**, p. 670]. This matrix appears in Figure 2.9.

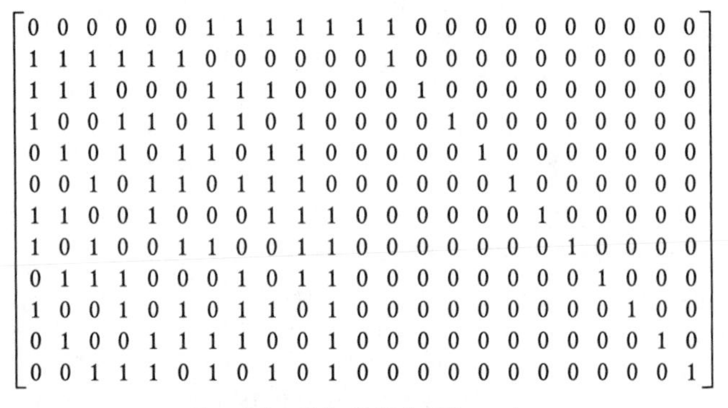

$$\begin{bmatrix}
0&0&0&0&0&0&1&1&1&1&1&1&1&0&0&0&0&0&0&0&0&0&0\\
1&1&1&1&1&1&0&0&0&0&0&0&1&0&0&0&0&0&0&0&0&0&0\\
1&1&1&0&0&0&1&1&1&0&0&0&0&1&0&0&0&0&0&0&0&0&0\\
1&0&0&1&1&0&1&1&0&1&0&0&0&0&1&0&0&0&0&0&0&0&0\\
0&1&0&1&0&1&1&0&1&1&0&0&0&0&0&1&0&0&0&0&0&0&0\\
0&0&1&0&1&1&0&1&1&1&0&0&0&0&0&0&1&0&0&0&0&0&0\\
1&1&0&0&1&0&0&0&1&1&1&0&0&0&0&0&0&1&0&0&0&0&0\\
1&0&1&0&0&1&1&0&0&1&1&0&0&0&0&0&0&0&1&0&0&0&0\\
0&1&1&1&0&0&0&1&0&1&1&0&0&0&0&0&0&0&0&1&0&0&0\\
1&0&0&1&0&1&0&1&1&0&1&0&0&0&0&0&0&0&0&0&1&0&0\\
0&1&0&0&1&1&1&1&0&0&1&0&0&0&0&0&0&0&0&0&0&1&0\\
0&0&1&1&1&0&1&0&1&0&1&0&0&0&0&0&0&0&0&0&0&0&1
\end{bmatrix}$$

FIG. 2.9. Paige's 12 by 23 matrix.

But neither Golay nor Paige gave a proof that his respective matrix had the required minimum Hamming distance between vectors in the row space. The laborious checking process was left to the reader. Leech wanted one where the minimum Hamming distance could be established easily—"I wanted a perspicuous proof, and this, in the light of my very limited knowledge of coding theory at that time, had to be technically elementary." [**76**] Note that if he had used either Golay's (Figure 1.23) or Paige's matrix (Figure 2.9), and formed a 12 by 24 matrix with the addition of a single overall parity check digit, then any vector in the row space of the new matrix would have an even number of 1's. Thus it would have at least eight 1's since the minimum number of 1's in any row generated by the old matrix was seven and adjoining an overall parity check digit can only increase this number.

Before continuing with Leech's construction, we wish to briefly consider the history of Paige's paper. In a sense this is an epilogue to Chapter 1 because it is another approach to the codes of both Hamming and Golay (but done independently five years after Hamming's paper of 1950).

To call the matrix in Figure 2.9 "Paige's matrix" is not quite accurate. It actually arose out of work done by E. H. Spanier in the summer of 1955 in connection with a classified government project called SCAMP at U.C.L.A. [**116**], [**98**, p. 3]. In this 1957 paper Paige credited Spanier with much of the work [**96**, p. 15].

At the beginning of the SCAMP project, several mathematicians proposed a set of problems for examination. Spanier, who had come to U.C.L.A. from the University of Chicago for the summer, and Paige, based at U.C.L.A., both participated in the project along with some twenty other mathematicians. But they were not working on the same problem [**97**], [**116**]. Spanier was focusing on "Substantialism," which he describes as follows:

> We consider n-tuples of elements from the field GF(2) (each element is to be regarded as either 0 or 1). The problem is to find a [small] set of such n-tuples such that every one of the 2^n possible n-tuples differs in at most one place from some n-tuple in the set. A set of n-tuples with these properties will be said to be a covering. [**114**, p. 1]

He credits I. J. Good with coining the term *substantialism*. This term suggests that the covering is "substantially all the n-tuples" since any n-tuple is within one coordinate of some element of the covering [**116**]. For example, if a combination lock could be opened even though one of the numbers dialed is incorrect, what is the smallest number of combinations that must be tried to guarantee the lock will open? (This is not unlike generalizations, described by Stein in [**117**, p. 458], to a soccer-pool problem first investigated by O. Taussky and J. Todd in 1948 [**119**].)

The first SCAMP working paper by Spanier dealt with minimal coverings in dimensions up to seven [**114**]. The second considered linear coverings; the third explored the existence of exact coverings and was titled "Economical Substantialism Mod 2" [**98**], [**115**]. In the course of the last two papers Spanier established the existence of the perfect Hamming binary codes, the two-word binary ϵ-error-correcting codes of length $2\epsilon + 1$ and both special Golay codes without being aware of the work of Golay or Hamming [**116**]. Not only does Paige's matrix first appear there, but so does the matrix in Figure 2.10 [**98**, p. 3], [**115**, p. 8].

$$\begin{bmatrix} 1 & 0 & 0 & 0 & 0 & 1 & 1 & 1 & 1 & 1 & 0 \\ 0 & 1 & 0 & 0 & 0 & 1 & 1 & 2 & 2 & 0 & 1 \\ 0 & 0 & 1 & 0 & 0 & 1 & 2 & 1 & 0 & 2 & 1 \\ 0 & 0 & 0 & 1 & 0 & 1 & 2 & 0 & 2 & 1 & 2 \\ 0 & 0 & 0 & 0 & 1 & 1 & 0 & 2 & 1 & 2 & 2 \end{bmatrix}$$

FIG. 2.10. Spanier's 5 by 11 matrix.

This matrix is a check matrix for the (11, 6) Golay code. (Cf. Figures 1.24 and 1.25.)

The rows of Paige's 12 by 23 matrix are a basis for the orthogonal complement to the subspace generated by the rows of the following 11 by 23 matrix (see Figure 2.11).

$$\begin{bmatrix}
1&0&0&0&0&0&0&0&0&0&0&1&1&1&1&0&0&1&1&0&1&0&0\\
0&1&0&0&0&0&0&0&0&0&0&1&1&1&0&1&0&1&0&1&0&1&0\\
0&0&1&0&0&0&0&0&0&0&0&1&1&1&0&0&1&0&1&1&0&0&1\\
0&0&0&1&0&0&0&0&0&0&0&1&1&0&1&1&0&0&0&1&1&0&1\\
0&0&0&0&1&0&0&0&0&0&0&1&1&0&1&0&1&1&0&0&0&1&1\\
0&0&0&0&0&1&0&0&0&0&0&1&1&0&0&1&1&0&1&0&1&1&0\\
0&0&0&0&0&0&1&0&0&0&0&1&0&1&1&1&0&0&1&0&0&1&1\\
0&0&0&0&0&0&0&1&0&0&0&1&0&1&1&0&1&0&0&1&1&1&0\\
0&0&0&0&0&0&0&0&1&0&0&1&0&1&0&1&1&1&0&0&1&0&1\\
0&0&0&0&0&0&0&0&0&1&0&1&0&0&1&1&1&1&1&0&0&0\\
0&0&0&0&0&0&0&0&0&0&1&1&0&0&0&0&0&1&1&1&1&1
\end{bmatrix}$$

FIG. 2.11. 11 by 23.

This matrix first appeared in the last SCAMP paper and then in Paige's 1957 paper [**98**, p. 3], [**78**, p. 16]. The correctness of Paige's matrix (Figure 2.9) was proved by showing that no six columns of the matrix in Figure 2.11 are linearly independent. This is demonstrated as follows: If some vector in the row space of Paige's matrix has fewer than seven 1's, then, since that vector's inner product with any row of the matrix in 2.11 is zero, the columns in the positions of those 1's would be linearly dependent, a contradiction. Neither Spanier nor Paige could recall exactly how the matrix in 2.11 was constructed [**97**], [**116**]. Of the verification process, Spanier said:

> What happened was that I played with this problem for a while and I came up with that matrix [Figure 2.11] and I said, 'Gee, how can I verify that?' I told Paige about it and he programmed it for the computer (SWAC). By the time it was done on the machine, I had checked it for myself. But I wasn't too certain about my own checking. [**116**]

Besides the existence of the perfect Hamming codes and the matrices generating the two perfect Golay codes, the last SCAMP working paper included a short, elementary proof that no perfect 2-error-correcting binary (90, 78) code exists —"there is no economical 2-dense subset of F^{90}." [**98**, p. 2] This proof, which we will now give, was credited to C. B. Tompkins.

Recall that the equality

$$1 + 90 + \binom{90}{2} = 2^{12}$$

indicates that such a code might exist. Golay in 1949 had shown that the code could not be linear and provided the nonexistence proof in the nonlinear case in his 1958 paper with a reference to S. K. Zaremba. The proof of Tompkins is much shorter.

Assume that such a code exists and that, without loss of generality, it contains the all-zero codeword. Each 90-tuple with exactly three 1's must be in a Hamming sphere of radius two centered on a codeword which has exactly five 1's, three of which are in the same coordinates as those of the 90-tuple. Consider in particular the set of eighty-eight 90-tuples with 1's in the first two coordinates and one 1 elsewhere. Each member of this set must be within two units of a unique codeword. Each such codeword has 1's in the first two positions and three 1's elsewhere. Two such codewords cannot have 1's in the same coordinate positions other than the first two, otherwise some 90-tuple in the above set would be within two units of two codewords. This means that the eighty-eight coordinate positions other than the first two must be divided into equinumerous disjoint sets of three. Since three does not divide eighty-eight, this is impossible.

Section 4. The Matrix for Leech's First Packing

Leech started with the symmetric matrix shown in Figure 2.12.

$$\begin{bmatrix}
0 & 0 & 0 & 0 & 0 & 0 & 0 & 0 & 0 & 0 & 0 & 0 \\
0 & 1 & 0 & 1 & 0 & 0 & 0 & 1 & 1 & 1 & 0 & 1 \\
0 & 0 & 1 & 0 & 0 & 0 & 1 & 1 & 1 & 0 & 1 & 1 \\
0 & 1 & 0 & 0 & 0 & 1 & 1 & 1 & 0 & 1 & 1 & 0 \\
0 & 0 & 0 & 0 & 1 & 1 & 1 & 0 & 1 & 1 & 0 & 1 \\
0 & 0 & 0 & 1 & 1 & 1 & 0 & 1 & 1 & 0 & 1 & 0 \\
0 & 0 & 1 & 1 & 1 & 0 & 1 & 1 & 0 & 1 & 0 & 0 \\
0 & 1 & 1 & 1 & 0 & 1 & 1 & 0 & 1 & 0 & 0 & 0 \\
0 & 1 & 1 & 0 & 1 & 1 & 0 & 1 & 0 & 0 & 0 & 1 \\
0 & 1 & 0 & 1 & 1 & 0 & 1 & 0 & 0 & 0 & 1 & 1 \\
0 & 0 & 1 & 1 & 0 & 1 & 0 & 0 & 0 & 1 & 1 & 1 \\
0 & 1 & 1 & 0 & 1 & 0 & 0 & 0 & 1 & 1 & 1 & 0
\end{bmatrix}$$

FIG. 2.12. The 12 by 12 matrix A_{12}.

Its entries are defined by $a_{ij} = 0$ if $i = 1$ or $j = 1$ or if $i + j - 4$ is a quadratic residue modulo 11, and $a_{ij} = 1$ otherwise [**72**, p. 665, 666]. Recall that the integer $d \neq 0$ is a quadratic residue modulo m if the congruence $x^2 \equiv d(\text{mod } m)$ has a solution. The quadratic residues of 11 are 1, 3, 4, 5 and 9 (and any integer congruent to them modulo 11). This matrix resembles one in Paley's paper (the source of A_8) which was defined as follows: $a_{ij} = 1$ if $i = 1$ or $j = 1$ or if $j - i$ is a quadratic residue of 11 and $a_{ij} = -1$ otherwise [**99**, p. 312]. Leech replaced the 1's and -1's of Paley's matrix with 0's and 1's, respectively, and then rearranged the rows to obtain a symmetric matrix. For example, since $j - 2 = j +$

$2 - 4$, the second rows correspond to each other. However, since $j - 3 \equiv j + 8 = j + 12 - 4$, the third row of Paley's matrix corresponds to the twelfth row of Leech's and so on.

Paley's matrix has the (Hadamard) property that the inner product, over Z, of any two distinct rows is zero (*orthogonal*) [**99**, p. 311]. In fact, the inner product of any two of its rows is the sum of exactly six 1's and six -1's, the 1's arising where the entries in the rows are the same and the -1's from where the rows differ. Thus any two rows of Leech's matrix A_{12} differ in exactly six places.

Initially, Leech generated a packing in E^{12} by taking all points with coordinates congruent modulo 2 to the coordinates of a row of either A_{12} or $\overline{A}_{12}$ and whose coordinates sum to a multiple of four [**72**, p. 666]. (The reader can compare this construction with that of Leech's packing in E^{16}, p. 78.) He then showed that the vector space generated by the row spaces of A_{12} and $\overline{A}_{12}$ modulo 2 is the lattice in E^{12} of integer points whose coordinate sum is even [**72**, p. 666, 667]. The rows of A_{12} and $\overline{A}_{12}$ correspond to the vertices of a cross polytope β_{12} inscribed in a 12-cube γ_{12}. (A *cross polytope*, β_n, in E^n, is a convex polytope which has as vertices the $2n$ endpoints of n equal-length mutually perpendicular line segments all of whose midpoints coincide. For example, β_1 is a line segment, β_2 a square and β_3 a regular octahedron.)

In order to use this matrix in E^{24}, Leech replaced its first row and first column with 1's except for the top left corner element and called the resulting matrix C_{12}, which is found in Figure 2.13. The matrix C_{12} has several useful properties. First, the vector sum modulo 2 of all the rows of C_{12} is the "all-1" row. This is easy to see since all the columns of this symmetric matrix have an odd number of 1's. Next, the presence of the all-1 vector in the row space of C_{12} makes the row space of C_{12} the same as that generated by the combined row spaces of A_{12} and $\overline{A}_{12}$. For example, the second row of A_{12} can be formed by summing the all-1 vector with the first two

$$\begin{bmatrix}
0 & 1 & 1 & 1 & 1 & 1 & 1 & 1 & 1 & 1 & 1 & 1 \\
1 & 1 & 0 & 1 & 0 & 0 & 0 & 1 & 1 & 1 & 0 & 1 \\
1 & 0 & 1 & 0 & 0 & 0 & 1 & 1 & 1 & 0 & 1 & 1 \\
1 & 1 & 0 & 0 & 0 & 1 & 1 & 1 & 0 & 1 & 1 & 0 \\
1 & 0 & 0 & 0 & 1 & 1 & 1 & 0 & 1 & 1 & 0 & 1 \\
1 & 0 & 0 & 1 & 1 & 1 & 0 & 1 & 1 & 0 & 1 & 0 \\
1 & 0 & 1 & 1 & 1 & 0 & 1 & 1 & 0 & 1 & 0 & 0 \\
1 & 1 & 1 & 1 & 0 & 1 & 1 & 0 & 1 & 0 & 0 & 0 \\
1 & 1 & 1 & 0 & 1 & 1 & 0 & 1 & 0 & 0 & 0 & 1 \\
1 & 1 & 0 & 1 & 1 & 0 & 1 & 0 & 0 & 0 & 1 & 1 \\
1 & 0 & 1 & 1 & 0 & 1 & 0 & 0 & 0 & 1 & 1 & 1 \\
1 & 1 & 1 & 0 & 1 & 0 & 0 & 0 & 1 & 1 & 1 & 0
\end{bmatrix}$$

FIG. 2.13. The 12 by 12 matrix C_{12}.

rows of C_{12} (see Figures 2.12 and 2.13). Another important result is that $C_{12}{}^2 \equiv I_{12}$ and (mod 2).

Leech formed the generating matrix for the lattice in E^{24} by placing I_{12} to the left of C_{12}. This 12 by 24 matrix (Figure 2.14), which will be denoted by $C = I_{12} \vdots C_{12}$, will be shown to have the necessary Hamming distance properties.

$$\begin{bmatrix}
1 & 0 & 0 & 0 & 0 & 0 & 0 & 0 & 0 & 0 & 0 & 0 & 0 & 1 & 1 & 1 & 1 & 1 & 1 & 1 & 1 & 1 & 1 & 1 \\
0 & 1 & 0 & 0 & 0 & 0 & 0 & 0 & 0 & 0 & 0 & 0 & 1 & 1 & 0 & 1 & 0 & 0 & 0 & 1 & 1 & 1 & 0 & 1 \\
0 & 0 & 1 & 0 & 0 & 0 & 0 & 0 & 0 & 0 & 0 & 0 & 1 & 0 & 1 & 0 & 0 & 0 & 1 & 1 & 1 & 0 & 1 & 1 \\
0 & 0 & 0 & 1 & 0 & 0 & 0 & 0 & 0 & 0 & 0 & 0 & 1 & 1 & 0 & 0 & 0 & 1 & 1 & 1 & 0 & 1 & 1 & 0 \\
0 & 0 & 0 & 0 & 1 & 0 & 0 & 0 & 0 & 0 & 0 & 0 & 1 & 0 & 0 & 0 & 1 & 1 & 1 & 0 & 1 & 1 & 0 & 1 \\
0 & 0 & 0 & 0 & 0 & 1 & 0 & 0 & 0 & 0 & 0 & 0 & 1 & 0 & 0 & 1 & 1 & 1 & 0 & 1 & 1 & 0 & 1 & 0 \\
0 & 0 & 0 & 0 & 0 & 0 & 1 & 0 & 0 & 0 & 0 & 0 & 1 & 0 & 1 & 1 & 1 & 0 & 1 & 1 & 0 & 1 & 0 & 0 \\
0 & 0 & 0 & 0 & 0 & 0 & 0 & 1 & 0 & 0 & 0 & 0 & 1 & 1 & 1 & 1 & 0 & 1 & 1 & 0 & 1 & 0 & 0 & 0 \\
0 & 0 & 0 & 0 & 0 & 0 & 0 & 0 & 1 & 0 & 0 & 0 & 1 & 1 & 1 & 0 & 1 & 1 & 0 & 1 & 0 & 0 & 0 & 1 \\
0 & 0 & 0 & 0 & 0 & 0 & 0 & 0 & 0 & 1 & 0 & 0 & 1 & 1 & 0 & 1 & 1 & 0 & 1 & 0 & 0 & 0 & 1 & 1 \\
0 & 0 & 0 & 0 & 0 & 0 & 0 & 0 & 0 & 0 & 1 & 0 & 1 & 0 & 1 & 1 & 0 & 1 & 0 & 0 & 0 & 1 & 1 & 1 \\
0 & 0 & 0 & 0 & 0 & 0 & 0 & 0 & 0 & 0 & 0 & 1 & 1 & 1 & 1 & 0 & 1 & 0 & 0 & 0 & 1 & 1 & 1 & 0
\end{bmatrix}$$

FIG. 2.14. The 12 by 24 matrix C.

The discussion follows closely that in Leech's paper [**72**, pp. 670, 671].

To show that any two vectors in the row space of C differ in at least eight places, it is sufficient to show that any such vector has at least eight 1's. This will be accomplished by showing that every row having at most three 1's in its first twelve places or in its last twelve places has at least eight 1's in all.

Rows having a single 1 in the first twelve places are rows of C which have eight 1's in total except for the first row which has twelve 1's. Rows having two 1's in their first twelve places are sums of pairs of rows of C. If one of the pair is the first row, the sum will have a 1 in the thirteenth place and five 1's thereafter for a total of eight 1's, since each row after the first has five 0's after the thirteenth place. If not, then the sum has 1's in six places in the last twelve entries where the rows differ. The total number of 1's is again eight.

Rows having three 1's in their first twelve places are sums of rows of C, one of which may be the first row while the other two certainly are not. Since the sum of two rows other than the first has six 1's and five 0's in the last eleven places, this sum, when added to the first row, yields five 1's in the last eleven places for a total of eight 1's in all. We have yet to deal with the case in which all three rows differ from the first.

Earlier in the paper Leech had considered the following 6 by 12 matrix D:

$$\begin{bmatrix} 1 & 0 & 0 & 0 & 0 & 0 & 0 & 1 & 1 & 1 & 1 & 1 \\ 0 & -1 & 0 & 0 & 0 & 0 & 1 & 0 & 1 & -1 & -1 & 1 \\ 0 & 0 & -1 & 0 & 0 & 0 & 1 & 1 & 0 & 1 & -1 & -1 \\ 0 & 0 & 0 & -1 & 0 & 0 & 1 & -1 & 1 & 0 & 1 & -1 \\ 0 & 0 & 0 & 0 & -1 & 0 & 1 & -1 & -1 & 1 & 0 & 1 \\ 0 & 0 & 0 & 0 & 0 & -1 & 1 & 1 & -1 & -1 & 1 & 0 \end{bmatrix}$$

FIG. 2.15. Leech's matrix D.

Leech formed the matrix D from two other matrices J_6 and D_6 by abutting them as $J_6 \vdots D_6$. The elements of D_6 are $d_{ii} = 0$, $d_{ij} = 1$ if i or j is 1 but not both, otherwise, d_{ij} is the quadratic residue symbol $\left(\frac{i-j}{5}\right)$, where $\left(\frac{i-j}{5}\right)$ is 1 if $i - j$ is a quadratic residue of 5 or -1 if $i - j$ is a quadratic nonresidue of 5. The matrix J_6 is formed from I_6 by changing the signs so as to make the row sums of D divisible by three. It is easy to check that each row of D as well as sums and differences of pairs of rows have six 0's. Since $D_6{}^2 \equiv -I_6 (\text{mod } 3)$, the rows of

$$D^* = \begin{bmatrix} 0 & 1 & 1 & 1 & 1 & 1 & 1 & 0 & 0 & 0 & 0 & 0 \\ 1 & 0 & -1 & 1 & 1 & -1 & 0 & -1 & 0 & 0 & 0 & 0 \\ 1 & -1 & 0 & -1 & 1 & 1 & 0 & 0 & -1 & 0 & 0 & 0 \\ 1 & 1 & -1 & 0 & -1 & 1 & 0 & 0 & 0 & -1 & 0 & 0 \\ 1 & 1 & 1 & -1 & 0 & -1 & 0 & 0 & 0 & 0 & -1 & 0 \\ 1 & -1 & 1 & 1 & -1 & 0 & 0 & 0 & 0 & 0 & 0 & -1 \end{bmatrix}$$

are linear combinations modulo 3 of those of D and vice versa. Similarly, the rows of D^* and the sums and differences of pairs of rows all have just six 0's. Leech then noted that any linear combination modulo 3 of rows of D having four or five 0's in either its first six elements or in its last six elements has just six 0's in all. Furthermore, as any other combination has at most three 0's in each of its first and last sets of six elements, it follows that every nonzero linear combination modulo 3 of rows of D has at most six 0's.

Incidentally, note that if we now delete any column of D, the remaining 6 by 11 matrix generates, modulo 3, the 3^6 vectors of the (11, 6) Golay ternary code. This follows because the vectors, which have at most six 0's, are at minimum distance five from each other. Any such (11, 6) ternary code is equivalent to the (11, 6) Golay code [**31**], [**101**].

Leech then used the rows of D to generate, modulo 3, the following matrix B_{12} consisting of only 0's and 1's:

$$\begin{bmatrix}
0&0&0&0&0&0&0&0&0&0&0&0\\
0&0&0&1&0&1&1&0&0&1&1&1\\
0&0&1&0&1&0&1&0&1&1&1&0\\
0&1&0&1&0&0&1&1&1&1&0&0\\
0&0&1&0&0&1&1&1&1&0&0&1\\
0&1&0&0&1&0&1&1&0&0&1&1\\
0&1&1&1&1&1&1&0&0&0&0&0\\
0&0&0&1&1&1&0&1&1&0&1&0\\
0&0&1&1&1&0&0&1&0&1&0&1\\
0&1&1&1&0&0&0&0&1&0&1&1\\
0&1&1&0&0&1&0&1&0&1&1&0\\
0&1&0&0&1&1&0&0&1&1&0&1
\end{bmatrix}$$

The matrix B_{12}.

Similarly, he generated $\overline{B_{12}}$, where as before $\overline{b_{ij}} = 1 - b_{ij}$. Since linear combinations of rows of D have at most six 0's, rows of $B_{12}, \overline{B_{12}}$ differ in at least six places and hence, like the rows of $A_{12}, \overline{A_{12}}$, they give the coordinates of the vertices of a cross polytope β_{12} inscribed in a cube γ_{12}. Thus, the matrices A_{12} and B_{12} differ only in the arrangement of rows and columns. (This fact depends upon the uniqueness of the equivalence class of Hadamard matrices of order 12 and is not in general valid for those of order 16 and upward, i.e., for those orders there are ways of inscribing cross polytopes in n-cubes so that the corresponding matrices are not row and column rearrangements of each other. Using B_{12} in place of A_{12} throughout avoids this argument.)

We are now ready for the case in which all three rows of C differ from the first. We wish to show that at most four of the six 1's in the last eleven places in the sum of two such rows correspond to 1's in the third. Assume that this is not the case. Thus the corresponding rows formed from A_{12} will have at least five 0's in common. Consider the case in which they have exactly five 0's in common. With suitable column permutation, we have the following situation:

first row	0	0	0	0	0	0	1	1	1	1	1	1
second row	0	0	0	1	1	1	0	0	0	1	1	1
mod 2 sum	0	0	0	1	1	1	1	1	1	0	0	0
third row	0	0	1	1	1	1	1	1	0	0	0	0

Because of the constraint on the 0's, the 1's in the third row must match either the first three or second three 1's in the modulo 2 sum of the first two rows. We have assumed the former. But then the following linear combination modulo 3 is in the row space of D:

first row	0	0	0	0	0	0	1	1	1	1	1	1
second row ×2	0	0	0	2	2	2	0	0	0	2	2	2
third row	0	0	1	1	1	1	1	1	0	0	0	0
mod 3 sum	0	0	1	0	0	0	2	2	1	0	0	0

This contradicts the fact that all nonzero vectors in the (12, 6) Golay code must have weight at least six. The other cases are treated similarly.

In essence, Leech used the extended Golay ternary code (the (12, 6) Golay code) to prove the correctness of the extended Golay binary code (the (24, 12) Golay code). Deleting any column of the generating matrices for the extended Golay codes yields matrices for the perfect Golay codes. Compare matrix D of Figure 2.15 with Figures 2.10, 1.24, 1.25, and 1.26. The latter three are check matrices for a (11, 6) Golay code.

This lengthy digression has shown that at most four of the six 1's in the last eleven places in the sum of two rows of C correspond to 1's in the third. This means that the sum of all three rows has at least four 1's in the last eleven places, a single 1 in the thirteenth place and three 1's in the first twelve places for a total of at least eight 1's in all. What all this bookkeeping shows is that no sum having three 1's in the first twelve places has less than eight 1's in all.

To deal with sums of rows having three or fewer 1's in the last twelve places, note that

$$C_{12} \cdot C = C_{12} \cdot I_{12} \vdots C_{12} = C_{12} \vdots I_{12},$$

and so the rows of $C^* = C_{12} \vdots I_{12}$ are linear combinations of the rows of C. This argument shows that any vector with three or fewer 1's in the last twelve places is a sum of three or fewer rows of C^*. From the structure of C^*, it is clear that the same arguments apply to it as apply to C. Thus any vector with three or fewer 1's in the last twelve places has at least eight 1's in all.

Any vector not already considered has at least four 1's in each half. Thus any vector in the row space of C, which we will denote by $R(C)$, has at least eight 1's. In passing, note that since the rows of C all have an even number of 1's, the number of 1's in any vector in $R(C)$ is even and at least eight.

In carrying out the above construction, Leech by-passed the formation of a 12 by 23 matrix with the appropriate distance properties and then the extension of its rows by overall parity checks. As we have already mentioned, by deleting any column of C, just such a 12 by 23 matrix (call it F) is formed, since the deletion process decreases the Hamming distance between any two vectors by at most one. This means that the minimum distance between any two vectors in the row space $R(F)$ of F is at least seven and these vectors form an exact 3-covering of the unit 23-cube.

Let a *neighborhood* of one of these vectors be a Hamming sphere of radius three centered on that vector. Since $R(F)$ is an exact 3-cover, any vertex of the unit 23-cube with exactly four 1's lies in a unique neighborhood of a vector in the row space with seven 1's. Furthermore, four of the seven 1's are in the same positions as the 1's in the vertex with exactly four 1's. Now identify any vertex with exactly four 1's with the set of four integers which are the coordinate positions of the four 1's. Call such a set a *4-set.* Do the same for those vectors with exactly seven 1's which are in $R(F)$. These are a family of

7-sets. Let $T = \{1, 2, 3, \ldots, 23\}$ be the set of column numbers of the matrix F. The family of 7-sets of T is an example of a Steiner system $S(4, 7, 23)$ which we will define in a moment. In a similar manner, the 8-sets of C form a Steiner system $S(5, 8, 24)$ and this fact will be used to help calculate the contact number of Leech's packing in E^{24}.

A Steiner system is a combinatorial system defined as follows. Let T be a set of r distinct objects and let p and q be positive integers. Any set with n elements will be called an *n-set.* A *Steiner system $S(p, q, r)$* is a collection of q-sets of T with the property that any p-set of T is contained in exactly one q-set of the collection. Nontrivial examples require $p < q < r$.

It is not known in general for which triplets p, q, r Steiner systems exist. However, for $p = 1, 2, 3, 4, 5$, nontrivial examples are known. For instance, consider $p = 1$. An $S(1, q, r)$ system is a family of disjoint q-sets whose union is an r-set. Clearly this exists if and only if r is a multiple of q. By letting $T = \{1, 2, 3, 4\}$ we see that one $S(1, 2, 4)$ system consists of the two set $\{1, 2\}, \{3, 4\}$.

In general, since there are $\binom{r}{p}$ possible p-sets and since each q-set contains $\binom{q}{p}$ distinct p-sets, there must be exactly $\binom{r}{p}/\binom{q}{p}$ q-sets in the family $S(p, q, r)$.

The finite projective plane with seven points provides a system $S(2, 3, 7)$.

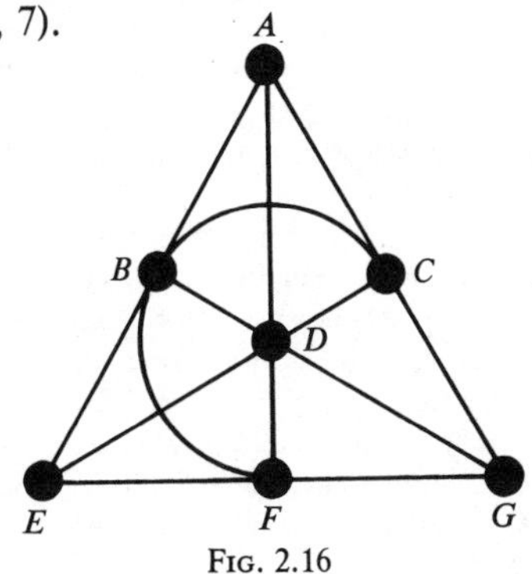

FIG. 2.16

The lines are:

$$\begin{array}{lll} ABE & BDG & CBF \\ ACG & CDE & EFG \\ ADF & & \end{array}$$

A *projective plane* consists of a set T whose elements are called *points*, and a family of subsets of T, called *lines*, which satisfy the following four axioms:

1. Two distinct points of T lie on one and only one line.
2. Any two lines meet in at least one point.
3. There exist three noncollinear points.
4. Every line contains at least three points. [**61**, p. 6]

The seven points and seven lines in Figure 2.16 satisfy these axioms. Since each line has three points, Axiom 1 implies that the lines (3-sets) form a Steiner system $S(2, 3, 7)$. For any prime t, such a plane, having $t^2 + t + 1$ points, can be constructed in which each line will have exactly $t + 1$ points. These planes are Steiner systems $S(2, t + 1, t^2 + t + 1)$. A bibliography and survey of Steiner systems can be found in [**34**].

For $p > 3$, known Steiner systems $S(p, q, r)$ are scarce. Only fourteen have been discovered: $S(5, 6, 12)$, $S(5, 8, 24)$ and their respective contractions* $S(4, 5, 11)$ and $S(4, 7, 23)$,

*A *contraction* of a Steiner system $S(p, q, r)$ is the Steiner system $S(p-1, q-1, r-1)$ formed as follows: Delete an element a from the r-set T. Pick all the q-sets of $S(p, q, r)$ *which contain* a and form corresponding $(q-1)$-sets by deleting a. These form a system $S(p-1, q-1, r-1)$. In fact, there is no known system $S(p, q, r)$ with $p = 4$ which is not the contraction of a known system $S(5, q + 1, r + 1)$ [**33**]. Unfortunately, this "step-down" process does not work in reverse; we cannot form a Steiner system $S(p + 1, q + 1, r + 1)$ from the system $S(p, q, r)$. However, it does make some nonexistence proofs possible. For example, if it is known that no Steiner system $S(p, q, r)$ exists for a certain triple (p, q, r), then neither does a Steiner system $S(p + 1, q + 1, r + 1)$ exist. This affords an even shorter proof of the nonexistence of an exact 2-cover in E^{90}. The 5-sets of such a cover would form a Steiner system $S(3, 5, 90)$ which would imply the existence of those of $S(2, 4, 89)$ and $S(1, 3, 88)$, but the latter is impossible since 3 does not divide 88.

together with $S(5, 6, 24)$, $S(5, 7, 28)$, $S(5, 6, 48)$, $S(5, 6, 72)$, $S(5, 6, 84)$ and their respective contractions [**33**]. The last ten of these were discovered only recently [**32**], [**92**]. The first four were shown by Ernst Witt, in 1938, to be unique up to any permutation of the elements of T [**126**], [**127**].

As we will see, they are intimately related to both the perfect (11, 6) and (23, 12) Golay codes. Moreover, as we will see in Chapter 3, they are also related to certain simple groups (groups with no nontrivial normal subgroups).

Recall that we have already formed a Steiner system $S(4, 7, 23)$. This is the family of vectors in the row space of the matrix F (the matrix C with one column deleted) which have exactly seven 1's. When each of these vectors is extended by an overall parity check, we get a family of vectors in E^{24} with eight 1's. We will now show that this family, which is the set of vectors in $R(C)$ with exactly eight 1's, forms a Steiner system $S(5, 8, 24)$.

Consider any vertex in the unit 24-cube with exactly five 1's. If one 1 is deleted in a certain coordinate and that coordinate is deleted in all the vertices of the unit 24-cube, the new vertex with exactly four 1's is a vertex of the unit 23-cube. The corresponding 4-set is a subset of T and lies in a unique 7-set of $S(4, 7, 23)$ which corresponds to a unique vector with seven 1's in $R(F)$. Thus, when the deleted coordinate is restored, the 5-set corresponding to the vertex in the unit 24-cube with exactly five 1's lies in a unique 8-set which corresponds to one of the vectors in $R(C)$ with exactly eight 1's. This means that the set of vectors in $R(C)$ with exactly eight 1's forms a Steiner system $S(5, 8, 24)$.

To obtain the first lattice packing in E^{24} given by Leech, take as sphere centers all points whose coordinates are congruent modulo 2 to vectors in $R(C)$ and have their sum divisible by four. One question remains—is this really a lattice packing? We present an easy proof.

Let L be the set of integer points both congruent modulo 2 to some vector on $R(C)$ and whose sum of coordinates is divisible by four. We must show that L is a group, that is, if $\mathbf{v}$ and $\mathbf{w}$ are lattice points, then so is $\mathbf{v} - \mathbf{w}$.

It is easy to see that the sum of the coordinates of $\mathbf{v} - \mathbf{w}$ is a multiple of four since the sum of two multiples of four is again a multiple of four. Suppose further that the coordinates of $\mathbf{v}$ and $\mathbf{w}$ are congruent modulo 2, respectively, to the coordinates of $\mathbf{c}$ and $\mathbf{d}$, which lie in $R(C)$. It is easy to show that the coordinates of $\mathbf{v} - \mathbf{w}$ are congruent modulo 2 to the coordinates of $\mathbf{c} - \mathbf{d}$, which is an element in $R(C)$. This shows that L is a lattice.

Let us now determine the contact number of the packing in E^{24} derived from this lattice. As before, we need to find (a) the minimum Euclidean distance between any two points of L and (b) the number of points of L at this distance from the origin.

By definition, elements of $R(C)$ with sixteen 0's and eight 1's belong to L. Furthermore, no point of L with an odd entry in a coordinate position can have fewer than eight nonzero coordinates since any such lattice point must have coordinates congruent modulo 2 to the coordinates of an element in $R(C)$ and no such element with a single 1 has fewer than eight 1's. Any lattice point with sixteen 0's and eight ± 1's which add up to a multiple of four has Euclidean norm $2\sqrt{2}$. Lattice points with all even coordinates and of minimum Euclidean norm have twenty-two 0's and two ± 2's. Their norm is also $2\sqrt{2}$. Thus the minimum distance between any two points of the lattice is $2\sqrt{2}$ and the packing is of spheres of radius $\sqrt{2}$.

To find the number of lattice points with sixteen 0's and eight ± 1's which add up to a multiple of four, we must first find the number of elements in $R(C)$ with sixteen 0's and eight 1's. This latter set of points has already been shown to form a Steiner system $S(5, 8, 24)$, which, by the formula on page 95 has $\binom{24}{5}/\binom{8}{5} = 759$ 8-sets. Since the eight ± 1's must add up to a multiple of four, there has to be an even number of 1's and -1's, yielding $(2^8/2)\cdot 759 = 97152$ lattice points of this type. Similarly, those having twenty-two 0's and two ± 2's comprise $2^2\cdot\binom{24}{2} = 1104$ points, making the contact number of this packing equal to 98256. The number of each possible shape is listed in Table 2.17.

TABLE 2.17

Shape	Number
$0^{16}, -1^8$	759
$0^{16}, -1^6, 1^2$	$759\cdot\binom{8}{2} = 21{,}252$
$0^{16}, -1^4, 1^4$	$759\cdot\binom{8}{4} = 53{,}130$
$0^{16}, -1^2, 1^6$	$759\cdot\binom{8}{2} = 21{,}252$
$0^{16}, 1^8$	759
$0^{22}, -2^2$	$\binom{24}{2} = 276$
$0^{22}, -2, 2$	$24\cdot 23 = 552$
$0^{22}, 2^2$	$\binom{24}{2} = 276$

Points closest to the origin in Leech's first packing.

Though it is impossible to sketch the points of L in twenty-four dimensions, Figure 2.5, p. 68, with coordinates doubled, represents a two-dimensional cross section. Of necessity it shows only lattice points whose last twenty-two coordinates are zero. The corresponding three-dimensional cross section is not hard to visualize.

The density of the packing generated by the lattice L is fairly easy to establish. Congruence modulo 2 to vectors in $R(C)$ means that 2^{12} in every 2^{24} points of E^{24} with integer coordinates are acceptable. The additional requirement that the row sum of any point must be a multiple of four reduces the acceptance to 2^{11} in every 2^{24} points.

Geometrically, this additional requirement has the effect of increasing the radii of spheres in the packing from 1 to $\sqrt{2}$, as happened, for example, in the packing of E^{16} (cf. p. 78).

Thus, on the average, there are $2^{11}/2^{24}$ lattice points per unit volume so that the density of this packing is

$$\frac{2^{11}}{2^{24}} \cdot \frac{\pi^{12} \cdot 2^{12}}{12!} = 0.0009647.$$

It is interesting to note that this density is about 39% of Rogers' upper bound for E^{24}, 0.002455.

The packing Leech was to give in his next paper (1967) in E^{24} achieved twice the density of this first packing and more than twice the contact number.

Section 5. The Leech Lattice

Soon after publishing his first 24-dimensional lattice packing L in 1964, Leech discovered yet denser packings in several Euclidean spaces, one of which was again E^{24}. The results amounted to a seventeen-page supplement to the 1964 paper, received by the editors June 30, 1965, and published in 1967 [**74**, p. 251].

The discussion of this new packing in E^{24}, whose lattice is known as the Leech lattice, $\Lambda = \Lambda_{24}$, would later become the star attraction of both papers. Its description occupied less than one page [**74**, pp. 262, 263]. Although Leech felt, [**73**], that it would probably be more important than the rest of the paper, he gave it no special prominence within the paper. The packing Λ was simply one of several improvements in the 1964 paper. Certainly at the time no one suspected that the lattice was to have a significant impact in the theory of finite simple groups, which had only recently awakened after being nearly dormant since the turn of the century [**37**], [**65**].

By the Jordan-Hölder theorem for finite groups, every finite group G has a composition series, that is, there is a sequence of subgroups extending from the identity element to G itself with each subgroup normal in the next and such that all corresponding quotient groups are simple. Furthermore, the collection of the quotient groups is unique up to reordering. It is possible, by a technique introduced by O. Schreier in 1926, to construct all possible multiplication tables for a group G from its composition factors [**65**, pp. 694, 695], [**107**]. By considering all possible lists of composition factors, we could then construct all possible groups. Thus a knowledge of the simple groups would lead to a knowledge of all possible finite groups. The problem of classifying finite groups is thus reduced to the classification of simple groups.

The theorem of Lagrange shows easily that all groups of prime order are simple. Another well-known infinite family of simple groups consists of the alternating groups, A_n, $n \neq 4$. Each is the subgroup of S_n consisting of all even permutations and has order $n!/2$. Several other infinite families are known [**65**, p. 708]. Besides these, up to 1965 there were five finite simple groups that did not fit into any infinite family. These had been discovered by Émile Mathieu in 1861 and 1873. Zvonimir Janko added to the Mathieu groups a sixth "sporadic" simple group in 1965 [**68**]. The list continued to grow and in 1968 John Conway used the Leech lattice, Λ, to generate three new sporadic simple groups. In August, 1980, the list of twenty-six entries (Appendix 6) was finally proved complete [**38**].

How did Leech find his new packing in E^{24}? One section in the first part of this new paper showed how the contact number of the packings in E^{2^m}, $m \geq 5$, could be decreased with no loss in density. For $m \geq 6$ he found a way to insert a new family of spheres in the old packing which, in effect, doubled its density [**74**, pp. 255, 256], [**77**].

Leech also discovered that a similar approach resulted in a doubling of both the density and the contact number of the original packing in E^{24} [**77**]. Since he made this discovery after submitting the 1967 paper (June 30, 1965), he included it with a covering letter, dated September 15, 1965. At that time he already had some notion of the importance of this new packing for he states in the covering letter:

> I enclose a new note which I hope it will be possible to add to my paper 'Notes on Sphere Packings.' This gives an account of a remarkable new lattice in 24 dimensions twice as dense as that of my former paper.... I sincerely hope it will be possible to include this note with the paper, as this would be a very suitable place for it and the result it embodies is probably more important than the rest of the paper. [**73**]

This addition necessitated Leech's rewriting the paper. The revision was received for publication on August 5, 1966, nearly a year later.

We now give Leech's definition of the new lattice. Each of the twenty-four coordinates of a lattice point will be expressed in binary notation with the digits named, from right to left, ones digit, twos digit, fours digit, etc. The binary numbers will satisfy the following conditions. First, the twenty-four ones digits of each lattice point are either all 0 or all 1. This means that a lattice point will have coordinates which are either all even or all odd. Second, the twos digits will form a row that is a vector in $R(C)$. Third, the fours digits will form rows that have even parity for those points whose ones digits are all 0 and have odd parity for those whose ones digits are all 1. The eights digits, sixteens digits, etc., are unrestricted.

Leech expressed negative integers in binary form using complementary representation, as follows [**72**, p. 662]. To represent a negative integer, first form the binary representation of its absolute value. Then find the binary complement of each digit and finally add 1. We illustrate the process in the case of the negative integer -3:

$\lvert -3 \rvert$	...0000011	(...indicates 0's)
Complement	...1111100	(...indicates 1's)
Complement $+1$	...1111101.	

Note that this binary representation of -3 added to the binary representation of 3 is indeed "zero," ...0000000.

This notation for the negative integers, known as the *twos complement* representation, is used extensively in computers [**10**, p. 60 ff].

Notice that the last three binary positions of -3 are 101, which is 5 in binary, and that $-3 \equiv 5 \pmod 8$. This holds for the other negative integers as well, that is, if b is negative and c is nonnegative and less than eight and if $b \equiv c \pmod 8$, then the last three binary digits of b form the binary representation of c.

Recall that for lattice points, the eights digits, sixteens digits, etc., of the binary representation of their coordinates are unrestricted. This means that the lattice consists of all vectors in E^{24} whose coordinates are congruent modulo 8 to vectors with coordinates between 0 and 7 which satisfy the three conditions given on the ones, twos and fours positions of their binary representations. In particular, the previous paragraph shows that a typical lattice point has either all even or all odd integer coordinates.

We should stop to check that the Leech lattice, Λ, is indeed a lattice. Now, Λ is twenty-four dimensional since it contains the twenty-four vectors, each having one coordinate 8 and all the rest 0, which are linearly independent. We must still establish that if $\mathbf{v}$, $\mathbf{w}$ are in Λ, then so is $\mathbf{v}-\mathbf{w}$.

Consider the binary representations of the coordinates of $\mathbf{v}$ and $\mathbf{w}$. For $\mathbf{v}$, the ones digits are either all 0 or all 1. Suppose that the ones digits of $\mathbf{w}$ are all 1. The negative of $\mathbf{w}$ is formed by complementing each of the digits in the binary representation of each coordinate entry and then adding 1 modulo 2. Since the ones digits are all 1, complementing makes them all zero and adding 1 makes them all 1 again. Thus the ones digits in $\mathbf{v}-\mathbf{w}$ are either all 0 or all 1. The twos digits of $\mathbf{w}$ form a vector in $R(C)$. Its complement is also in $R(C)$. Since the ones digits of $\mathbf{w}$ are all 1, the addition of 1, in the for-

mation of $-\mathbf{w}$, simply complements all the twos digits again. We again obtain a vector in $R(C)$. By hypothesis the fours digits of $\mathbf{w}$ form a row of odd parity. The complementing process keeps this parity odd and, by virtue of the twos positions forming a vector in $R(C)$, the adding of 1 to each position results in the adding of a row of even parity to the fours digits row which keeps the fours digits row odd. Suppose the ones digits of $\mathbf{v}$ are all 0. Then the ones digits of $\mathbf{v}-\mathbf{w}$ are all 1. Since the twos digits of both $\mathbf{v}$ and $-\mathbf{w}$ form vectors in $R(C)$, so do those of $\mathbf{v}-\mathbf{w}$. Furthermore, Lemma A2.1 shows that an even number of 1's are carried to the fours digits row of $\mathbf{v}-\mathbf{w}$. Also note that the fours digits of $\mathbf{v}$ form a row of even parity, so that the row of fours digits in $\mathbf{v}-\mathbf{w}$ has odd parity. Thus $\mathbf{v}-\mathbf{w}$ is in Λ. An example of these calculations appears in Figure 2.18.
It is an easy exercise to show that $\mathbf{v}-\mathbf{w} \in \Lambda$ no matter what the ones digits of $\mathbf{v}$ and $\mathbf{w}$ are. This shows that Λ is a lattice.

Leech's lattice Λ has the following alternate definition which will be of use in Chapter 3.

Let $\mathbf{c} = (c_1, c_2, \ldots, c_{24})$ be any vector in $R(C)$ and let m be an integer. Let $\mathbf{c}(m)$ denote the set of all integer vectors $\mathbf{v} = (v_1, v_2, \ldots, v_{24})$ in E^{24} such that

1. $\sum_{i=1}^{24} v_i = 4m$,
2. $v_i \equiv m \pmod 4$ if $c_i = 0$,
3. $v_i \equiv m+2 \pmod 4$ if $c_i = 1$.

Then Λ is the union of all the $\mathbf{c}(m)$. Showing the equivalence of the two definitions is an instructive exercise, which is left for the reader.

The contact number of the lattice packing formed by the lattice Λ is the number of lattice points of Λ at minimum distance from the origin. Because of the restrictions involving Λ, there are three basic types of such lattice points: those with sixteen 0's and eight ± 2's, those with twenty-two 0's and two

v:	(2,	4,	−4,	0,	0,	…,	0,	−2,	−2,	6,	6,	2,	2,	…,	2)
1's	0	0	0	0	0	…	0	0	0	0	0	0	0	…	0
2's	1	0	0	0	0	…	0	1	1	1	1	1	1	…	1
4's	0	1	1	0	0	…	0	1	1	1	1	0	0	…	0
8's	0	0	1	0	0	…	0	1	1	0	0	0	0	…	0
.	.	.	.	.	.		.	.	.	.	.	.	.		.
.	.	.	.	.	.		.	.	.	.	.	.	.		.
.	.	.	.	.	.		.	.	.	.	.	.	.		.

w:	(−3,	1,	1,	1,	1,	…,	1,	5,	1,	1,	1,	−3,	1,	…,	1)
1's	1	1	1	1	1	…	1	1	1	1	1	1	1	…	1
2's	0	0	0	0	0	…	0	0	0	0	0	0	0	…	0
4's	1	0	0	0	0	…	0	1	0	0	0	1	0	…	0
8's	1	0	0	0	0	…	0	0	0	0	0	1	0	…	0
.	.	.	.	.	.		.	.	.	.	.	.	.		.
.	.	.	.	.	.		.	.	.	.	.	.	.		.
.	.	.	.	.	.		.	.	.	.	.	.	.		.

−w:	(3,	−1,	−1,	−1,	−1,	…,	−1,	−5,	−1,	−1,	−1,	3,	−1,	…,	−1)
1's	1	1	1	1	1	…	1	1	1	1	1	1	1	…	1
2's	1	1	1	1	1	…	1	1	1	1	1	1	1	…	1
4's	0	1	1	1	1	…	1	0	1	1	1	0	1	…	1
8's	0	1	1	1	1	…	1	1	1	1	1	0	1	…	1
.	.	.	.	.	.		.	.	.	.	.	.	.		.
.	.	.	.	.	.		.	.	.	.	.	.	.		.
.	.	.	.	.	.		.	.	.	.	.	.	.		.

v − w:	(5,	3,	−5,	−1,	−1,	…,	−1,	−7,	−3,	5,	5,	5,	1,	…,	1)
1's	1	1	1	1	1	…	1	1	1	1	1	1	1	…	1
2's	0	1	1	1	1	…	1	1	0	0	0	0	0	…	0
4's	1	0	0	1	1	…	1	1	1	1	1	1	0	…	0
8's	0	0	1	1	1	…	1	0	1	0	0	0	0	…	0
16's	0	0	1	1	1	…	1	1	1	0	0	0	0	…	0
.	.	.	.	.	.		.	.	.	.	.	.	.		.
.	.	.	.	.	.		.	.	.	.	.	.	.		.
.	.	.	.	.	.		.	.	.	.	.	.	.		.

FIG. 2.18. A twos complement calculation.

± 4's, and those with twenty-three ± 1's and one ± 3. It is easy to check that points of each type are $4\sqrt{2}$ units from the origin regardless of the arrangement of the $+$'s and $-$'s.

Since the binary representation of -2 has a 1 in the fours digits place while that of 2 does not, any lattice point with sixteen 0's and eight ± 2's must have an even number of -2's. Similarly, those with twenty-two 0's and two ± 4's have no such restriction. Thus, the sub-lattice of Λ with even coordinates is the same as L but with coordinates doubled. The various shapes of the $97152 + 1104 = 98256$ such points are given in Table 2.19 (cf. Table 2.17).

TABLE 2.19

Shape	Number
$0^{16}, -2^8$	759
$0^{16}, -2^6, 2^2$	$759 \cdot \binom{8}{2} = 21{,}252$
$0^{16}, -2^4, 2^4$	$759 \cdot \binom{8}{4} = 53{,}130$
$0^{16}, -2^2, 2^6$	$759 \cdot \binom{8}{2} = 21{,}252$
$0^{16}, 2^8$	759
$0^{22}, -4^2$	$\binom{24}{2} = 276$
$0^{22}, -4, 4$	$24 \cdot 23 = 552$
$0^{22}, 4^2$	$\binom{24}{2} = 276$

Vectors in Λ, with even coordinates, at distance $4\sqrt{2}$ from $\mathbf{0}$.

Since the minimum distance of this packing is double that of the lattice packing L, the spheres have twice the radius as those in the lattice L. Hence, spheres centered on the points in Λ with even coordinates form a packing as dense as the previous one. But we have yet to consider the vectors in Λ with odd coordinates!

For convenience, the binary representations of ± 1 and ± 3 are listed below:

$$\begin{array}{rl} 1 & \ldots 0000001 \\ -1 & \ldots 1111111 \\ 3 & \ldots 0000011 \\ -3 & \ldots 1111101. \end{array} \tag{2.20}$$

These representations, together with the lattice restrictions on the fours digits place, show that closest points with odd coordinates have either an even number of -1's and one -3 (the -1's being in the places of the 1's of a vector of $R(C)$) or an odd number of -1's and one $+3$ (all of which are in the places of the 1's of a vector of $R(C)$). One can think of the ∓ 3 replacing a ± 1 in each of 24 possible locations. Thus, in total there are $2^{12} \cdot 24 = 98304$ such points. These are listed in Table 2.21.

TABLE 2.21

Shape	Number
$-1^{23}, 3$	24
$-1^{16}, 1^7, -3$	$759 \cdot 8 = 6{,}072$
$-1^{15}, 1^8, 3$	$759 \cdot 16 = 12{,}144$
$-1^{12}, 1^{11}, -3$	$2576 \cdot 12 = 30{,}912$
$-1^{11}, 1^{12}, 3$	$2576 \cdot 12 = 30{,}912$
$-1^8, 1^{15}, -3$	$759 \cdot 16 = 12{,}144$
$-1^7, 1^{16}, 3$	$759 \cdot 8 = 6{,}072$
$1^{23}, -3$	24

Vectors in Λ, with odd coordinates, at distance $4\sqrt{2}$ from $\mathbf{0}$.

This makes the contact number of Λ, $98256 + 98304 = 196560$, more than double that of the lattice packing L.

The density of this packing is an easy calculation. Consider the binary representations of the coordinates of a vector $\mathbf{v}_0$ with shape $(1^{23}, -3)$. Recall that the coordinates of any point of the lattice, when written in binary form, have the ones digits either all 0's or all 1's, the twos digits forming a vector in $R(C)$, and the fours digits forming a vector of even or odd parity according to whether the ones digits are either all 0's or all 1's. If $\mathbf{v}$ is in Λ and has even coordinates, then $\mathbf{v} + \mathbf{v}_0$ has odd coordinates and is in Λ. Similarly, any vector in Λ with odd coordinates can be written as $\mathbf{v} + \mathbf{v}_0$ where $\mathbf{v}$ is in Λ and has even coordinates. Thus Λ consists of the old lattice L (at double scale) plus its translate by $\mathbf{v}_0$ and hence the packing has density 0.001929, twice that of the previous packing. This is nearly 79% of Rogers' upper bound for E^{24}, 0.002455. (Table A1.1 in Appendix 1 compares the density of this packing with that of others in nearby dimensions.)

Although Leech's 1967 paper finishes the construction of Λ which Chapter 3 will use, Leech's work exploiting codes in sphere packing did not stop. N. J. A. Sloane wrote him in late 1969 to tell him how other coding theory results could be applied to sphere packings [**111**], [**75**]. Describing this joint effort, Leech remarked:

> My joint work with Sloane came about when he drew my attention to the use of BCH codes in 64 dimensions and of various codes in 9–15 dimensions. And then we were well away, aiding and abetting each other in the course of preparing our joint papers. [**77**]

Their work culminated in "Sphere Packings and Error-Correcting Codes" written in early 1971 [**83**].

However, the lattice packing Λ still stands as the densest known packing in E^{24}, lattice or not.

Not all questions about sphere packing in twenty-four dimensions have been answered. Is Leech's packing in E^{24} the densest possible? Does it have the highest possible contact number? Leech felt that ". . . one can hardly doubt that it is the densest possible packing in E^{24}," but offered no proof [**74**, p. 266]. Recently, A. M. Odlyzko and Sloane have shown that the packing has, in fact, the highest contact number possible in E^{24} [**95**], while E. Bannai and Sloane have shown that this local arrangement is unique [**9**].

We are now ready to see how this lattice led to the discovery of new simple groups.

CHAPTER 3

FROM SPHERE PACKING TO NEW SIMPLE GROUPS

Section 1. Is There an Interesting Group in Leech's Lattice?

What finite groups lie hidden in a lattice? For example, consider the standard integer lattice in E^2 pictured in Figure 3.1. Let G be the family of distance-preserving maps of the plane which fix the origin and carry the lattice into itself. These maps are usually called *isometries* or *Euclidean motions.* They need not preserve orientation. The group G has order eight, since (1, 0) must and can be mapped to one of the four points $(\pm 1, 0)$ or $(0, \pm 1)$, and then (0, 1) must and can be mapped to either of two of these points. Representative matrices for the group G are:

$$\begin{bmatrix} 1 & 0 \\ 0 & 1 \end{bmatrix}, \begin{bmatrix} 1 & 0 \\ 0 & -1 \end{bmatrix}, \begin{bmatrix} -1 & 0 \\ 0 & 1 \end{bmatrix}, \begin{bmatrix} -1 & 0 \\ 0 & -1 \end{bmatrix},$$

$$\begin{bmatrix} 0 & 1 \\ 1 & 0 \end{bmatrix}, \begin{bmatrix} 0 & 1 \\ -1 & 0 \end{bmatrix}, \begin{bmatrix} 0 & -1 \\ 1 & 0 \end{bmatrix}, \begin{bmatrix} 0 & -1 \\ -1 & 0 \end{bmatrix}.$$

Thus G is isomorphic to the dihedral group D_4, the group of symmetries of a square. Indeed, the isometries of the lattice coincide with the symmetries of the square whose vertices are the four points at distance 1 from the origin shown in Figure 3.1.

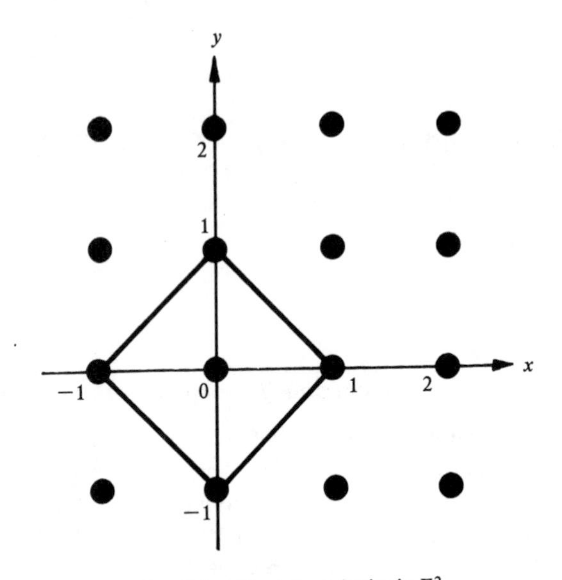

FIG. 3.1. The integer lattice in E^2.

One subgroup of the symmetry group G which is of interest is the subgroup of *rotations*. These are the orientation-preserving symmetries with determinant 1. In the case of G, this subgroup is cyclic of order four.

Leech was aware that the symmetry groups of his lattice might be of interest. Coxeter had written to him that "the symmetry groups of your new packings will provide several Ph.D. theses" in a letter dated May 18, 1963, referring to Leech's manuscript of his 1964 paper [**79**]. So the question became: "What symmetry groups are associated with the polytope, P, in E^{24} whose vertices are the 196560 points of the Leech lattice Λ_{24} which are closest to the origin?" In retrospect, Leech commented that "... I was alert to the possibilities early on, and probably thought about groups associated with the new lattice almost at once, but not very seriously until my 1967/68 sabbatical year." [**79**]

Leech's first task was to try to find the types of symmetries of the lattice. He already knew that the contact number was 196560, i.e., that any lattice point was adjacent to 196560 others at distance $4\sqrt{2}$, the minimum distance of the lattice. He found further that any pair of adjacent lattice points is adjacent to 4600 other lattice points. If one point of the pair is taken to be the origin, this implies that the other is a vertex of P and, further, that this vertex of P is adjacent to 4600 other vertices of P. There are basically three shapes of vertices to check:

$$(0^{16}, \pm 2^8), (0^{22}, \pm 4^2), (\pm 1^{23}, \pm 3).$$

We will carry out the calculations in just one case. Let $\mathbf{v}$ be the vertex $(4, 4, 0, 0, 0, \ldots, 0)$, which is adjacent to those vertices with one 4 in one of the first two positions, one ± 4 in the last twenty-two positions and 0's elsewhere. The total number of such adjacent vertices is $2 \cdot 2 \cdot 22 = 88$. The vertex $\mathbf{v}$ is also adjacent to those vertices with shape $(0^{16}, \pm 2^8)$ which have a 2 in each of the first two positions. Recall that the ± 2's are in the positions of an 8-set in $R(C)$. From Table A2.3, 77 8-sets contain the first two coordinate positions. Since each vertex with shape $(0^{16}, \pm 2^8)$ must have an even number of 2's and -2's, there are five free sign choices for the six remaining positions in the 8-set which yields a total of $77 \cdot 2^5 = 2464$ vertices with shape $(0^{16}, \pm 2^8)$ adjacent to $\mathbf{v}$. The last case consists of counting those vertices of shape $(\pm 1^{23}, \pm 3)$, if there are any, adjacent to $\mathbf{v}$. In any event, the first two positions of such a vertex must consist of a 1 and a 3. The paragraph preceding Table 2.21 shows that the first two positions of the corresponding vectors in $R(C)$ must differ (one will contain a 0, the other a 1). Since exactly half of the vectors in $R(C)$ do, the number of vertices with this shape adjacent to $\mathbf{v}$ is $2^{11} = 2048$, which makes a total of $88 + 2464 + 2048 = 4600$ adjacent to $\mathbf{v}$. This indicates that any vertex of P "looks like" any other vertex in the sense that they all have the same number of neighbors. Leech also found that any (equilateral) triangle of adjacent points of Λ is adjacent to 891 others, that any (regular) tetrahedron of adjacent lattice points is adjacent to 336 others and, finally, that any regular simplex of five lattice points is adjacent to 170 others. (See Appendix 3.) We mean that a pair of adjacent lattice points is adjacent to a third point if each point in the pair is at the minimum distance, $4\sqrt{2}$, from the third point, etc. When translated in terms of the polytope P, these conditions fall one step behind that in the lattice since the origin is always considered to be a member of the triangle, tetrahedron or simplex, i.e., any pair of adjacent vertices of P is adjacent to 891 others, etc. (In fact, the reader should be aware that we will continue to focus more on the polytope P than on the entire lattice Λ.)

On the basis of this information, Leech felt that "... it is evident (though not conclusively proved) that all such adjacent vertices, pairs of vertices, etc., are equivalent under symmetry operations of the whole configuration [P]..." [**79**]. In other words, he suspected that the symmetry group G of P has elements which send any vertex to any other (G is transitive in P), send any adjacent pair of vertices to any other adjacent pair (G is "locally" doubly transitive), etc.

A simple example in E^2 brings the ideas of the previous paragraph down to earth.

The best lattice packing in E^2, Figure 2.3, has contact number six. Figure 3.2 shows the six vertices closest to the origin. Note that these vertices form a regular hexagon and that any of these six vertices is at minimum distance from exactly two others of the six. This suggests that any vertex "looks like" any other in the hexagon (which is, of course, clear) and that the symmetry group, D_6, is transitive. However, D_6 is not only transitive but also sends any pair of adjacent vertices to any other pair of adjacent vertices.

From the numbers 196560, 4600, 891, 336 and 170 (for

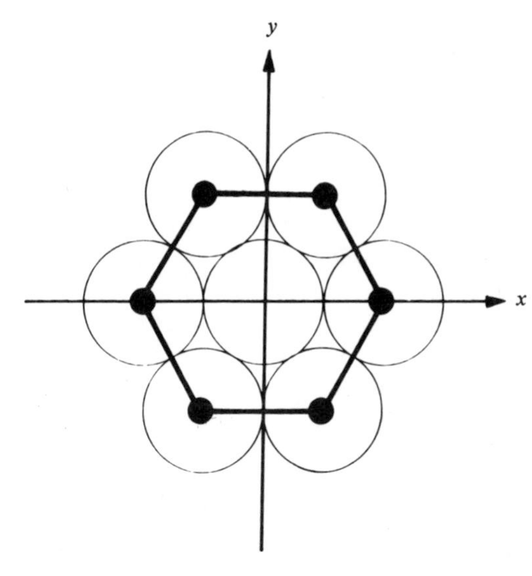

FIG. 3.2

geometric reasons 170 should be written as 10 + 160—see Appendix 3), and from the intimate relationship of the symmetry group of P with the Mathieu group M_{24} (which we shall define in Section 4), and the (24, 12) Golay code, Leech conjectured a lower bound on the order of the symmetry group of P. This number, the least common multiple of the two numbers $196560 \cdot 4600 \cdot 891 \cdot 336 \cdot 120$ and $2^{12} \cdot |M_{24}|$, is 8,315,553,613,086,720,000, or in factored form, $2^{22} \cdot 3^9 \cdot 5^4 \cdot 11 \cdot 13 \cdot 23$ [**80**].

We are not yet ready to look into the structure of the symmetry group of P, nor into how M_{24} and the (24, 12) Golay code are related to that group. These matters will be dealt with in Section 4.

At this point we wish to note that Leech, after having finished his two papers on sphere packings, turned to the problem of analyzing their respective symmetry groups. Of particular interest was the lattice Λ in E^{24}. His conjectured lower bound actually turned out to be the order of the symmetry group of P. But this is not the main point of interest. What drove him on was the suspicion that either the symmetry group of P or some very large subgroup or subgroups of the symmetry group were simple [**78**]. (Recall that new simple groups were being discovered at this time. We have already mentioned Janko's "sporadic" simple group, J.) Specifically, Leech felt that because of the multiple "local symmetry," "it was unlikely that any substantial subconfiguration could be fixed by a normal subgroup." [**78**] Let us illustrate what this means.

Consider the cube with the vertices painted alternately black and white—see Figure 3.3. The black vertices are connected to form a regular tetrahedron. The symmetry group of the cube has order $8 \cdot 3 \cdot 2 = 48$. Those symmetries which send one black vertex into another in fact take all black vertices to black vertices. Thus they fix the tetrahedron as a whole. It is easy to see that if b is a black-to-black symmetry and g a black-to-white symmetry, then $g^{-1}bg$ is a black-to-

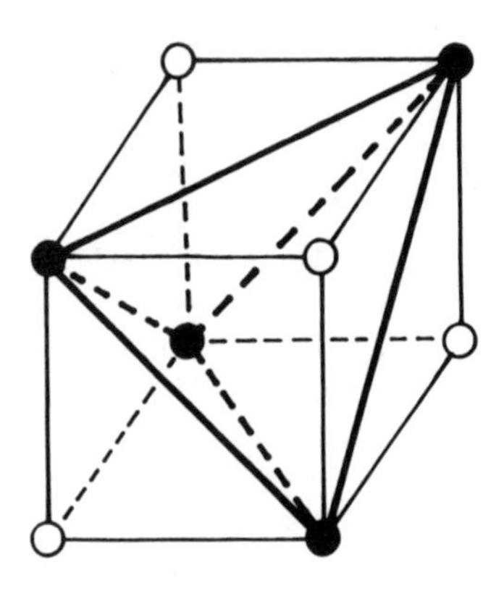

FIG. 3.3

black symmetry (and, of course, a white-to-white symmetry). This means that the subgroup of all black-to-black symmetries is normal in the whole symmetry group. (This fact could also be deduced from the fact that the order of the subgroup that takes black to black, which is $4 \cdot 3 \cdot 2 = 24$, is half that of the group of all symmetries of the cube.)

As a second example, consider the regular tetrahedron in Figure 3.4. The symmetry group of the tetrahedron is the symmetric group S_4 which has the alternating group A_4 as a normal subgroup. This latter group is, in fact, the group of rotations of the tetrahedron. If the midpoints of opposite edges are joined, the resulting line segments are mutually perpendicular and are fixed by the subgroup of A_4 isomorphic to the Klein 4-group: 1, $(AB)(CD)$, $(AC)(BD)$ and $(AD)(BC)$. This subgroup is of index three in A_4 and is normal in both A_4 and S_4. Other operations of A_4 permute the three line segments. Again, though no coloring scheme is possible, there is a "substantial subconfiguration" which is fixed by a normal subgroup.

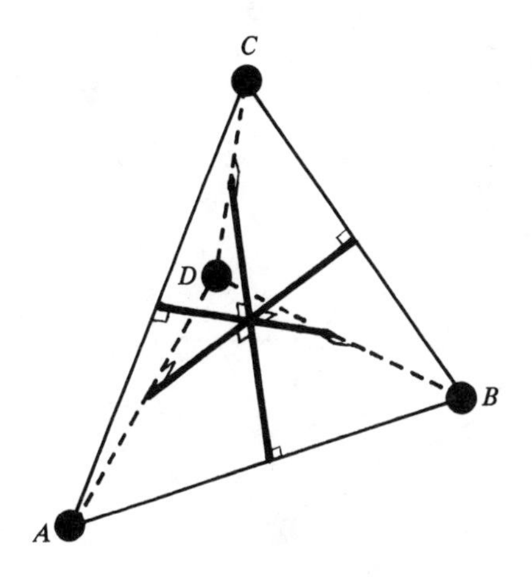

FIG. 3.4. The regular tetrahedron.

On the other hand, the regular icosahedron (Figure 3.5) has no such "substantial subconfiguration" which is fixed by a large normal subgroup of its symmetry group. Though this group is not simple, the rotation group, being A_5, is [**71**, pp. 17-19]. Note that the icosahedron has "local symmetry" up through pairs of adjacent vertices.

But experience in three dimensions is not enough. Leech recalled that

> The nearest analogue from which to guess in 24 dimensions was Gosset's lattice in 8 dimensions (5_{21} in Coxeter's notation, as in his *Regular Polytopes*, 1948, [**29**], which I knew of from soon after its publication). In this lattice, which is the set of all points in E^8 whose coordinates are integers, either all even or all odd, with sum divisible by four, each vertex is adjacent to 240 others, each pair to 56 others, each triad to 27 others, each tetrad to 16

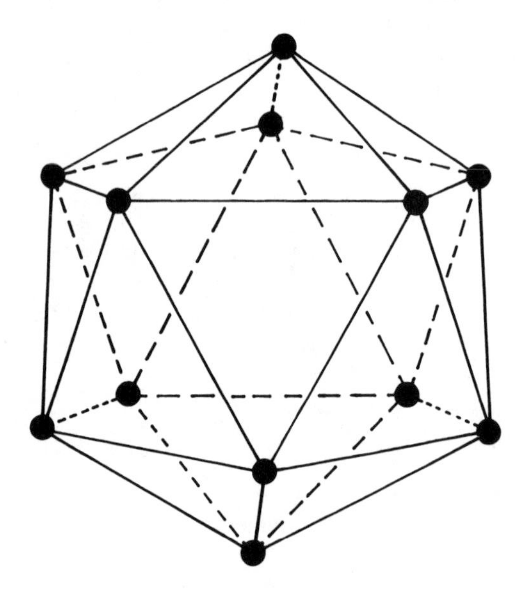

FIG. 3.5. The regular icosahedron.

> others, each pentad to 10 others, each hexad to 6 others (forming a triangular prism) and each heptad to $1 + 2$ others (obviously inequivalent in the prism). Multiplying up we get $192 \cdot 10!$ for the whole group including reflections, $96 \cdot 10!$ for the normal subgroup of direct rotations only, and $48 \cdot 10!$ for the simple group on factoring out the reflection in the origin (a rotation in even dimensions) and considering the 120 diameters. As you will see, the analogy is very close.... Experience with regular figures in 3 or more (particularly 4) dimensions may have helped, but this configuration was the main analogue. [**81**]

Further, Leech adds:

> I would not claim to have established that there was no central subconfiguration—it merely seemed too unlikely to be worth considering once one has rejected the obvious sorts of possibilities. The lack of a coloring scheme was one of these straightforward considerations. [The conviction of the existence of a large simple group] grew on me as I convinced myself with more and more cases that the symmetries were transitive up to pentads of adjacent vertices. But I hadn't formally proved even that! Indeed, I had really done so little that I could only offer it as being something which I was confident would repay full investigation. The main point is that in configurations of this nature (at any rate within my experience) if there are normal subgroups then one can pick out 'central subconfigurations' which they fix, and that if things are too symmetrical, then you can't pick these out. But this is all very heuristic; without aspiring to the standards of either, it's more like Ramanujan's ideas than Hardy's. Ramanujan was always full of bright ideas, some of them provable, a few not, a few just plain wrong, but always fascinating. [**81**]

Leech did not prove his conjectures. Looking back, he observed,

> My skill at group theory was insufficient for me to get anywhere near to solving this problem. [**78**]

Unable to solve the problem, he tried to interest others in it. The story of this effort at salesmanship does not appear in the published papers. Because it helps to show how mathematics evolves, and in particular, how the critical leap that takes an idea from one field to another occurs, it deserves to be told in detail.

Section 2. The Hard Sell of a Simple Group

Leech left Glasgow University in October, 1967, to spend a year as a research fellow at the Atlas Computer Laboratory near Harwell, England [**78**], [**89**]. Apparently, the Laboratory had the policy of having about half a dozen mathematicians around doing basic research [**89**].

While there, Leech renewed his acquaintance with John McKay, then another of the research fellows. Though Leech and McKay were not working on the same problems, McKay soon became acquainted with Leech's lattice and the fact that Leech suspected a large simple group was present in its symmetries. At the time McKay was working on the existence of a sporadic simple group of order $2^7 \cdot 3^5 \cdot 5 \cdot 17 \cdot 19 =$ 50,232,960, predicted by Z. Janko, later to be known as J_3 [**89**], [**65**, pp. 706, 709]. (The existence of J_3 was established in 1968 [**65**, p. 709].)

Leech wrote to Coxeter to find either a student of Coxeter's or someone else who might tackle the problem. Coxeter replied, saying that he had no graduate student who was capable of dealing with it [**89**]. A little later, still while at the Atlas Laboratory, Leech and McKay attended weekly lectures at nearby Oxford University. Of those visits McKay recalls:

> ... Leech and I both pestered several people, including Graham Higman, to have a look at this thing, to find the automorphism [symmetry] group, but nobody seemed terribly interested. [**89**]

At the time Higman was also trying to establish the existence of J_3.

But the situation was not to remain static very long. McKay went over to Cambridge University to discuss J_3. He recalled that either the day he returned from Cambridge, or the day before, he mentioned to John Thompson and John Conway, who were at Cambridge, that the symmetry group of Leech's lattice might indeed be interesting [**89**]. Conway was interested but did not work on the problem for a few months. As Leech acknowledged later [**78**], it was not easy to get someone to work on his question: "I dangled the problem under various noses, including those of Coxeter, Todd, and Graham Higman; but Conway was the first to swallow the bait—he had it from McKay visiting Cambridge."

With his first efforts, Conway managed to determine the size of the symmetry group—

$$2^{22}\cdot 3^{9}\cdot 5^{4}\cdot 7\cdot 11\cdot 13\cdot 23 = 8{,}315{,}553{,}613{,}086{,}720{,}000.$$

This group he called $\cdot 0$ (pronounced "dotto"). This confirmed Leech's conjecture, although Conway was unaware that Leech had conjectured the order of the group or even a lower bound for it [**26**], [**79**]. The group $\cdot 0$ is not simple, but the quotient of $\cdot 0$ by its two-element center turns out to be simple, as we shall see in Section 5. This quotient group is denoted $\cdot 1$, which is pronounced "dot one." During his investigations, Conway also discovered two other new sporadic simple groups, $\cdot 2$ and $\cdot 3$, of orders

$$2^{18}\cdot 3^{6}\cdot 5^{3}\cdot 7\cdot 11\cdot 23 = 42{,}305{,}421{,}312{,}000$$

and

$$2^{10}\cdot 3^{7}\cdot 5^{3}\cdot 7\cdot 11\cdot 23 = 495{,}766{,}656{,}000,$$

respectively. These will be defined at the end of this chapter in Section 5.

His results appeared in the *Proceedings of the National Academy of Sciences* (*U.S.A.*), communicated by Saunders Mac Lane on August 15, 1968, [**23**], less than a year after Leech had gone to the Atlas Laboratory. That brief two-and-one-half page announcement concludes with these acknowledgements:

> I should like above all to thank John Leech for discovering his Lattice and so freely disseminating information about it, John McKay for bringing it to my notice, and Professor John Thompson for stimulating conversations that began on the day the group was discovered and have not yet ceased. [**23**, p. 400]

The details followed in [**24**] and still later as part of a text on finite simple groups [**25**].

Section 3. Twelve Hours on Saturday, Six on Wednesday

Conway had always been somewhat interested in groups even though he specialized in combinatorics. Some of his experience with simple groups had been gained while verifying Janko's assertions about his new sporadic simple group, J [**26**]. This group is generated by the two 7 by 7 matrices in Figure 3.6 whose entries are from the field GF(11) [**68**], [**69**].

$$\begin{bmatrix} 0 & 1 & 0 & 0 & 0 & 0 & 0 \\ 0 & 0 & 1 & 0 & 0 & 0 & 0 \\ 0 & 0 & 0 & 1 & 0 & 0 & 0 \\ 0 & 0 & 0 & 0 & 1 & 0 & 0 \\ 0 & 0 & 0 & 0 & 0 & 1 & 0 \\ 0 & 0 & 0 & 0 & 0 & 0 & 1 \\ 1 & 0 & 0 & 0 & 0 & 0 & 0 \end{bmatrix} \begin{bmatrix} -3 & 2 & -1 & -1 & -3 & -1 & -3 \\ -2 & 1 & 1 & 3 & 1 & 3 & 3 \\ -1 & -1 & -3 & -1 & -3 & -3 & 2 \\ -1 & -3 & -1 & -3 & -3 & 2 & -1 \\ -3 & -1 & -3 & -3 & 2 & -1 & -1 \\ 1 & 3 & 3 & -2 & 1 & 1 & 3 \\ 3 & 3 & -2 & 1 & 1 & 3 & 1 \end{bmatrix}$$

FIG. 3.6. The two generators of Z. Janko's group J.

When McKay contacted Conway at Oxford and told him about Leech's lattice and that Leech thought there was an interesting (large, simple) group present, Conway was intrigued.

> I got interested, and as I looked at the thing, it became obvious to me that there really was a [large, simple] group involved. [**26**]

At this point Conway went to see Thompson:

> I was a bit in awe of him actually, because he was the best group theorist in the world, and everybody knew that. And I thought he

> was a very serious person. Anyhow, I told him once or twice that I thought this thing was interesting. [**26**]

Although Thompson expressed some curiosity, he did nothing immediately. Conway felt that

> . . . people were sort of coming up to him [Thompson] every other day and saying, 'Here, I've got an interesting idea, why don't you investigate it?' [**26**]

Later, when Conway confronted Thompson with this allegation, Thompson did not deny it. But then, according to Conway, Thompson said, "If you work out the order of this group, I will start to be interested." [**26**]

Conway accepted Thompson's remark as a challenge. He planned to work on the problem for twelve hours every Saturday, from noon to midnight, and for six hours every Wednesday, from six in the evening until midnight. Furthermore, he was prepared to stay at this regimen for several months, if necessary, until the problem was settled:

> On the first Saturday, I had a last cup of coffee and everything, kissed the wife and kids good-by, went and locked myself in the front room, and started to work. [**26**]

However, instead of months, the project required only hours.

Conway started by writing down everything he knew about the problem on a large roll of white paper, resembling a roll of old-fashioned butcher's paper. He realized that the group involved would be the group of symmetries of the vertices of the convex hull of those lattice points closest to the origin, i.e., the symmetry group of *P*. At this point he carried out the calculations much as Leech had done—finding the number of vertices of *P* (which is the number of spheres touching the sphere centered at the origin), the number of vertices of *P* at minimum distance from any fixed vertex of *P* and so on up to the number of vertices of *P* at minimum distance from any four-dimensional simplex of five adjacent points of *P*.

As he carried out these calculations, he found in each case

that the numbers were independent of the choice of adjacent pair, triangle, etc. Like Leech, he became convinced that the symmetry group was transitive on vertices, pairs of adjacent vertices, triangles whose vertices were all at minimum distance from each other, and so on. From this he deduced that the order of the group, if his assumptions concerning transitivity were correct, was either a certain number or twice it [**26**]. With this information at hand, Conway contacted Thompson.

> It was about six in the evening. I telephoned him and dictated this order to him saying that it was either this number or twice it. And then he got really terribly excited and a short time later telephoned me back with the correct order of the group and the fact that there were new simple groups associated with it— ·1, ·2 and ·3. (We used to joke that, if you wanted a new group and if you dreamt of an order, all you had to do was to pick up a telephone, dial 61200, ask for John Thompson, and then dictate the integer. The results could be quite spectacular.) But he hadn't been able to do the crucial thing and prove it existed [give a set of generators—in this case a set of 24 by 24 matrices], which was my job. [**26**]

Conway then went back to work trying to find a set of generators for the symmetry group.

> At this point I knew, at least in principle, that I could compute a matrix that had to be in the group. I started writing down a 24 by 24 matrix, filling in the entries piece by piece. And it wasn't entirely a routine calculation. But I couldn't quite see how to prove that the group generated by this matrix, and some others, was the group I was interested in.
>
> Anyway, I telephoned Thompson again and told him that I had this matrix, but that I was feeling quite exhausted (even though it was only ten o'clock) and was going to bed. I would talk more about it tomorrow. Then I hung the telephone up.
>
> Then I thought, 'No, I won't. I'll just see if I can at least see, in principle, how to do it.' The problem was to show that a spanning set [which will be defined later] would be sent by this matrix to another spanning set. For some reason I must have had a blind

> spot. Anyway, it suddenly dawned on me as soon as I had finished telephoning him the second time that I was being stupid. So I wrote down a list of forty vectors such that if the matrix fixed those as a whole, then it was all right. And I checked one of them. [**26**]

Conway immediately called Thompson again.

> ... I said, 'Well now, look, I've done a fortieth of this thing and I know it's all going to work and so, really now, I'm going to bed. [**26**]

But Conway didn't go to bed.

> I just said, 'Well, how bloody stupid to give up,' and so I carried on. At a quarter past midnight, I telephoned Thompson again, saying that it was all done. This group is there. It was absolutely fantastic—twelve hours had changed my life. Especially since I had envisioned it going on for months—every three days spending six or twelve hours on the damned thing. [**26**]

Conway went over to Thompson's department office the next day, Sunday, and worked on and talked about the group all day. These discussions continued all week [**26**].

Section 4. The Structure of $\cdot 0$

A. Introduction. This section follows Conway's work in [**24**] closely and supplies many omitted details, although occasionally in a different notation. Algebraic notation, not the functional notation of Chapter 1, will be used since much of the work involves permutations, which multiply more conveniently from left to right.

First, recall that $\cdot 0$ is the group of symmetries of P. This is the same as the family of Euclidean congruences in E^{24} which preserve the origin and fix the Leech lattice as a whole. Conway calls these symmetries rotations, because as it turns out, each can be represented by a 24 by 24 orthogonal matrix whose determinant is 1.

Let $\Omega = \{0, 1, 2, 3, \ldots, 22, \infty\}$. Think of the elements of Ω

as the residue classes of the integers modulo 23, together with the symbol ∞. We will use Ω to index the coordinates in E^{24}, for it is conveniently related to the construction of M_{24}, the Mathieu group on 24 letters, which will be discussed later in this section (pp. 127–137).

Let

$$\begin{aligned}
\mathbf{v}_0 &= (1, 0, 0, \ldots, 0, 0)\\
\mathbf{v}_1 &= (0, 1, 0, \ldots, 0, 0)\\
\mathbf{v}_2 &= (0, 0, 1, \ldots, 0, 0)\\
&\vdots\\
\mathbf{v}_{22} &= (0, 0, 0, \ldots, 1, 0)\\
\mathbf{v}_\infty &= (0, 0, 0, \ldots, 0, 1).
\end{aligned}$$

Any vector $\mathbf{v}$ of E^{24} has the form

$$\mathbf{v} = \sum_{i \in \Omega} x_i \mathbf{v}_i$$

where the x_i's are real. For each nonempty subset S of Ω, denote the vector

$$\sum_{i \in S} \mathbf{v}_i$$

by the symbol $\mathbf{v}_S$.

Vectors $\mathbf{w}$ of the form $8\mathbf{v}_i$ have already been shown to be in the Leech lattice. When operated on by a rotation (an element of $\cdot 0$), each of these vectors is mapped to a lattice point at a distance 8 from the origin. This image vector is also the vector formed by multiplying by 8 each entry in the ith ($i \in \Omega$) column of the matrix which represents the rotation. Thus each entry of the matrix is rational and has a denominator divisible by eight [**25**, p. 237].

B. More on Λ. Instead of Leech's definition of his lattice, Λ, Conway used the equivalent one given in page 104, which, in the new notation, goes as follows:

Let **c** be any vector in $R(C)$, the row space of Leech's generating matrix, and m an integer. Let $\mathbf{c}(m)$ denote all integer vectors $\mathbf{v}$ in E^{24} such that

1. $\sum_{i \in \Omega} x_i = 4m$,

2. $x_i \equiv m \pmod 4$ if $c_i = 0$,

3. $x_i \equiv m + 2 \pmod 4$ if $c_i = 1$.

Then Λ is the union of all the $\mathbf{c}(m)$ for all $m \in Z$.

Conway showed in [**24**, p. 80] that Λ is generated by linear combinations over Z of the 759 vectors of the form $2\mathbf{v}_K$ where K is an 8-set in $R(C)$* together with one vector with odd coordinates, say $\mathbf{v}_\Omega - 4\mathbf{v}_\infty = (1, 1, 1, \ldots, 1, -3)$. This is accomplished by first letting X, Y and Z be the sublattices spanned by all vectors of the respective forms $2\mathbf{v}_K$ (K an 8-set in $R(C)$), $4\mathbf{v}_T$ (T, a 4-set of Ω, i.e., $\mathbf{v}_T$ is a 24-tuple with four 1's and twenty 0's) and $4\mathbf{v}_i - 4\mathbf{v}_j$ ($i, j \in \Omega$) over the integers. Next we show that $X \supset Y$. Let $T = T_0$ be any 4-set of Ω. Corollary A2.5 (of Appendix 2) shows that T is contained in exactly five 8-sets which we will designate by $T + T_1, \ldots, T + T_5$ since any two such 8-sets intersect in exactly T. Thus $T_i + T_j = (T + T_i) + (T + T_j)$ so that any two members of $\{T_0, T_1, \ldots, T_5\}$ form an 8-set. If T, U and V are any three distinct members of this set, we have

$$4\mathbf{v}_T = 2\mathbf{v}_{T+U} + 2\mathbf{v}_{T+V} - 2\mathbf{v}_{U+V}$$

is in X so that $Y \subset X$.

It is also easy to show that $Y \subset Z$. Pick two 4-sets T, U, of Ω

---

*To say that K is in $R(C)$ is not exactly true. However, since $R(C)$ is generated by a Steiner system $S(5, 8, 24)$, and all such systems are equivalent up to permutation of the coordinate entries, we shall continue to use this designation.

which have three elements in common and suppose i is in T but not U and j is in U but not T. Then

$$4\mathbf{v}_i - 4\mathbf{v}_j = 4\mathbf{v}_T - 4\mathbf{v}_U.$$

Thus, since $Z = \mathbf{0}(0)$, $Y = \mathbf{0}(4m)$, $m \in Z$. This follows because $Y \supset Z$ and Y contains a vector from $\mathbf{0}(4)$. But then X contains all the vectors in the sets $\mathbf{c}(4m)$ where $\mathbf{c}$ is in $R(C)$, since $Y \subset X$ and the 8-sets span $R(C)$. This means that X, together with the one vector $\mathbf{v}_\Omega - 4\mathbf{v}_\infty$ with odd coordinates, generates all vectors in the sets $\mathbf{c}(m)$, m in Z, so that the 760 vectors generate Λ.

For any two vectors in this generating set it is easy to show that $\mathbf{v} \cdot \mathbf{v} \equiv 0 \pmod{16}$ and $\mathbf{v} \cdot \mathbf{w} \equiv 0 \pmod{8}$. This means that the same two congruences hold for any two vectors in Λ. Conway then defined Λ_n as the set of all vectors $\mathbf{v}$ in Λ such that $\mathbf{v} \cdot \mathbf{v} = 16n$. This is the set of all points in Λ at a distance $4\sqrt{n}$ from the origin.

The set Λ_1 is empty since no point of the lattice Λ other than the origin is closer than a distance $4\sqrt{2}$ from the origin.

The set Λ_2 consists of the 196560 lattice points closest to the origin, which are the vertices of P. From Tables 2.19 and 2.21, these are conveniently divided into three disjoint sets, Λ_2^2, Λ_2^3, Λ_2^4, depending upon their shape, as shown in Table 3.7.

TABLE 3.7

Set	Shape	Number
Λ_2^2	$0^{16}, \pm 2^8$	$759 \cdot 2^7 = 97{,}152$
Λ_2^3	$\pm 1^{23}, \pm 3$	$24 \cdot 2^{12} = 98{,}304$
Λ_2^4	$0^{22}, \pm 4^2$	$\binom{24}{2} \cdot 2^2 = 1{,}104$

The subsets of Λ_2

For example, Λ_2^2 consists of those vectors with eight ± 2's—an even number of -2's and 2's in each vector.

Thus

$$\Lambda_2 = \Lambda_2^2 \cup \Lambda_2^3 \cup \Lambda_2^4.$$

Λ_3 is used in the definition of $\cdot 3$, the third simple group Conway discovered (cf. Section 5).

C. The Mathieu Groups. Associated with any Steiner system $S(p, q, r)$ is its *automorphism group*, that is, the group of permutations of the r elements which permute the q-sets among themselves. Of particular interest to us is $S(5, 8, 24)$ and its automorphism group, the Mathieu group M_{24}. Since $S(5, 8, 24)$ is unique up to permutation of the symbols of the 24-set, there is only one group M_{24}.

Later in this section we shall define M_{24} as the subgroup of S_{24} generated by a certain set of three permutations. This definition will be shown to be equivalent to the one given above, with most of the proof appearing in Appendix 4.

The Mathieu group M_{24} is simple [**21**, p. 283]; indeed it is one of the first five sporadic simple groups discovered [**65**, p. 709]. It has another important property, which first drew attention to it [**86**], [**87**], its transitivity.

A permutation group on a k-set is *n-ply transitive* if, for any two sequences of n distinct elements of the k-set, x_1, $x_2, \ldots, x_n$ and $y_1, y_2, \ldots, y_n$, there is an element ϕ of the group such that $(x_i)\phi = y_i$, $i = 1, 2, \ldots, n$. For instance, the symmetric group on k letters, S_k, is n-ply transitive for each n, $1 \leq n \leq k$. Note that a k-ply transitive group is necessarily $(k - 1)$-ply, for $k \geq 2$. The group M_{24} is quintuply (5-ply) transitive. This will be shown later in this section.

The group M_{23} is defined to be the subgroup of M_{24} which fixes one point in the 24-set (the one-point* *stabilizer*

---

*Later, we shall have occasion to speak of a group of permutations of an n-set which fix some subset of the n-set *as a whole.* This means that the elements of the subset may be permuted by the subgroup and is different from the group of permutations which fix that subset element-wise, i.e., fix each point within the subset.

in M_{24}). It is equivalent to the automorphism group of a Steiner system $S(4, 7, 23)$, the contraction of a Steiner system $S(5, 8, 24)$.

In a similar fashion, M_{22} is defined to be the one-point stabilizer in M_{23}. However, it is of index 2 in its automorphism group [**48**, p. 77].

The Mathieu group M_{12} is defined as the automorphism group of a Steiner system $S(5, 6, 12)$, and M_{11} is its one-point stabilizer. The group M_{11} is also the automorphism group of a Steiner system $S(4, 5, 11)$.

Along with M_{24}, the other four Mathieu groups are simple, and together they form the first five sporadic simple groups. While M_{12} is quintuply transitive, M_{23} and M_{11} are quadruply transitive [**21**, pp. 151, 164, 165], and M_{22} is triply transitive. Other than the symmetric groups and the alternating groups, the four groups M_{11}, M_{12}, M_{23} and M_{24} are the only known quadruply transitive groups, and M_{12} and M_{24} are the only known quintuply transitive permutation groups on any finite set [**65**, p. 703]. No permutation groups other than S_n, $n \geq 6$, A_n, $n \geq 7$ are known to be sextuply transitive.

We now construct M_{24}, which will play a central role in the analysis of $\cdot 0$.

The *general linear group* $GL_2(23)$ is the group of all one-to-one linear maps of the two-dimensional vector space over the finite field GF(23). We may think of it as the group of all invertible 2 by 2 matrices with entries from GF(23). The *special linear group* $SL_2(23)$ consists of the elements of $GL_2(23)$ of determinant 1. Note that $SL_2(23)$ is a normal subgroup of $GL_2(23)$. The *projective special linear group* $PSL_2(23)$ is the group formed by factoring $SL_2(23)$ by its center. It is also known as the "linear fractional group" and is sometimes denoted LF(2, 23) or $L_2(23)$.

By considering the matrices that commute with the particular matrices

$$\begin{bmatrix} 1 & 1 \\ 0 & 1 \end{bmatrix} \quad \text{and} \quad \begin{bmatrix} 1 & 0 \\ 1 & 1 \end{bmatrix},$$

it is easy to show that the center of $SL_2(23)$ consists of just the two matrices of the form:

$$\begin{bmatrix} k & 0 \\ 0 & k \end{bmatrix}, \qquad k^2 = 1,$$

namely,

$$\begin{bmatrix} 1 & 0 \\ 0 & 1 \end{bmatrix} \quad \text{and} \quad \begin{bmatrix} -1 & 0 \\ 0 & -1 \end{bmatrix}.$$

This means that $PSL_2(23)$ can be thought of as the set of matrices of the form

$$\begin{bmatrix} a & b \\ c & d \end{bmatrix}$$

that satisfy $ad - bc \equiv 1 \pmod{23}$, where we identify the matrices

$$\begin{bmatrix} a & b \\ c & d \end{bmatrix} \quad \text{and} \quad \begin{bmatrix} -a & -b \\ -c & -d \end{bmatrix}.$$

We will also need some notions from projective geometry.

Let us recall the definitions of the real projective plane and the real projective line. A projective plane was defined earlier as a set T, whose elements are called points, together with a family of subsets of T, called lines, which satisfy these axioms:

1. Two distinct points of T lie on one and only one line.
2. Any two lines meet in at least one point.
3. There exist three noncollinear points.
4. Every line contains at least three points.

Let F be any field and consider the family of nonzero ordered triples of elements from F. Call two such ordered triples, (a, b, c) and (a', b', c') *equivalent* if there is a $k \neq 0$ in F such that $(a', b', c') = (ka, kb, kc)$. (It is easy to check that this describes an equivalence relation.) The family of equivalence classes of such ordered triples constitute the *points* of a projective plane. A *line* in this plane is the set of nonzero linear combinations over F of two given points. The coordinates of any single ordered triple which lies in a given point are the *homogeneous coordinates* for that point. Thus, both (2, 0, 5) and (4, 0, 10) describe the same point. When F is the field of real numbers, we obtain the *real projective plane*. When F is a finite field, we obtain a *finite projective plane*. For example, in case F is GF(2), the projective plane with seven points is obtained. (Compare Figures 3.8 and 2.16.)

In an analogous manner, using ordered pairs from a field instead of ordered triplets, we can construct the *projective line*. The field of real numbers yields the *real projective line* while any finite field yields a *finite projective line*. Note that if the field has, say, n elements, $x_1, x_2, \ldots, x_n$, the line has

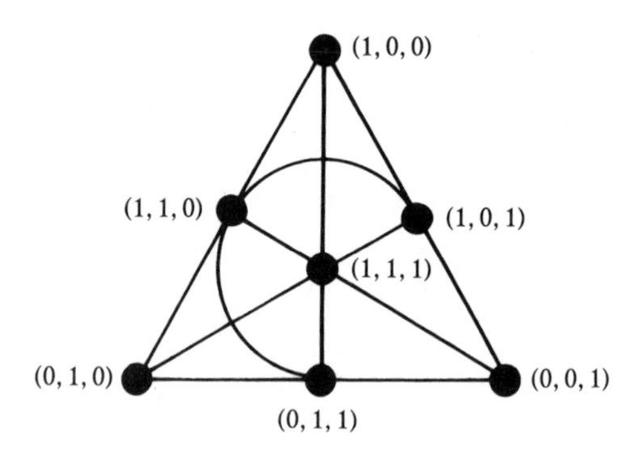

FIG. 3.8. The projective plane with seven points.

$n + 1$ points, represented by the elements $(1, x_1)$, $(1, x_2)$, $\ldots$, $(1, x_n)$ and $(0, 1)$.

For the present we are interested in the projective plane formed from the finite field, $F = \mathrm{GF}(23)$, and in particular, the projective line PL(23). Now, the ordered pairs (0, 1), (0, 2), (0, 3), ..., (0, 22) all belong to the same equivalence class, which we represent by the symbol 0. Similarly, "1" represents the equivalence class containing the pairs (1, 1), (2, 2), (3, 3), ..., (22, 22), since each quotient is 1. In short, we represent the point containing (m, n), $n \neq 0$, by the element m/n in GF(23). The symbol ∞ will represent the point consisting of those ordered pairs, (1, 0), (2, 0), ..., (22, 0), whose second entry is zero. Note that each point (equivalence class) has 22 members. We identify PL(23) with the set $\Omega = (0, 1, 2, \ldots, 22, \infty)$ mentioned earlier. For instance, 5 denotes the equivalence class or point, $\{(5, 1), (10, 2), (15, 3), (20, 4), (2, 5), \ldots, (18, 22)\}$.

The group $\mathrm{PSL}_2(23)$ acts in a natural way on the set Ω. In fact, $\mathrm{PSL}_2(23)$ can be thought of as the set of permutations of Ω of the form

$$x \mapsto \frac{ax + b}{cx + d}$$

where a, b, c, and d are in GF(23) and $ad - bc = 1$. To see this, let $\left[\begin{smallmatrix} a & b \\ c & d \end{smallmatrix}\right]$ be an element of $\mathrm{SL}_2(23)$, $x \in \Omega$, and (x_1, x_2) be the equivalence denoted by its common ratio. The equation

$$\begin{bmatrix} a & b \\ c & d \end{bmatrix} \begin{bmatrix} x_1 \\ x_2 \end{bmatrix} = \begin{bmatrix} ax_1 + bx_2 \\ cx_1 + dx_2 \end{bmatrix}$$

shows that the point x is taken to the point

$$\frac{ax_1 + bx_2}{cx_1 + dx_2},$$

which is in the same equivalence class as

$$\frac{a\dfrac{x_1}{x_2}+b}{c\dfrac{x_1}{x_2}+d},$$

and is represented by some element of Ω. The reader may check that the mapping defined on Ω is in fact a permutation of Ω.

We will now show that $PSL_2(23)$ is generated by the three permutations

$$\alpha\colon\ x \mapsto x+1$$

$$\beta\colon\ x \mapsto 2x$$

$$\gamma\colon\ x \mapsto -1/x.$$

First, note that these permutations are in $PSL_2(23)$ since

$$x+1 = \frac{1\cdot x + 1}{0\cdot x + 1},$$

$$2x = \frac{18\cdot x + 0}{0\cdot x + 9},$$

and

$$-1/x = \frac{0\cdot x - 1}{1\cdot x + 0}.$$

(The fact that $x \mapsto 2x$ is in $PSL_2(23)$ is equivalent to the assertion that 2 is a quadratic residue modulo 23.) In each case the appropriate 2 by 2 determinant has value 1. (This explains why we chose 9 and 18 rather than, say, 1 and 2, for the second matrix.)

Now note that

$$\beta^s\alpha^r\colon\ x \mapsto 2^s x + r$$

and

$$\beta^s\alpha^r\gamma\alpha^t\colon \quad x \mapsto t + \frac{-1}{2^s x + r}.$$

The powers of 2 form the multiplicative group 1, 2, 3, 4, 6, 8, 9, 12, 13, 16, 18, which is the set of eleven quadratic residues modulo 23. Since 2^s takes on $(23-1)/2$ values and r any value in GF(23), there are $(23-1)/2\cdot 23$ permutations of Ω of the form

$$x \mapsto 2^s x + r,$$

an instance of the form

$$x \mapsto \frac{ax+b}{cx+d}$$

with $c = 0$ and $a \neq 0$. Similarly, since t can be any element of GF(23), there are $23\cdot 23\cdot(23-1)/2$ permutations of the form

$$x \mapsto t + \frac{-1}{2^s x + r},$$

which is again of the form

$$x \mapsto \frac{ax+b}{cx+d}$$

but with $c \neq 0$. All together, α, β and γ generate

$$\frac{(23-1)}{2}\cdot 23\cdot 24 = 6072$$

permutations of Ω.

The number of solutions of $ad-bc = 1$, with a, b, c, d, in GF(23), is the number of elements in $SL_2(23)$. To compute how many such solutions there are, first observe that the ordered pair (a, b) can be chosen in $23^2 - 1$ ways. The ordered pair (c, d) can be chosen in $23^2 - 23$ ways so as not to be a linear multiple of (a, b). Hence, there are $(23^2-1)\cdot(23^2-23)$ 2 by 2 matrices

$$\begin{bmatrix} a & b \\ c & d \end{bmatrix}$$

with entries in GF(23) of nonzero determinant. Of these,

$$(23^2-1)\cdot(23^2-23)/22 = 12144$$

have determinant 1. Hence, $PSL_2(23)$ has $12144/2 = 6072$ elements. Consequently, α, β and γ generate all of $PSL_2(23)$.

Actually, $PSL_2(23)$ is generated by α and γ alone, for it is easily shown that $\beta = \alpha^5\gamma\alpha^5\gamma\alpha^{14}\gamma\alpha^{18}$. The element β is included to simplify later calculations. The cycle structures of α, β and γ are given in Figure 3.9.

$\alpha = (\infty)(0\ 1\ 2\ 3\ \ldots\ 22)$

$\beta = (\infty)(0)(1\ 2\ 4\ 8\ 16\ 9\ 18\ 13\ 3\ 6\ 12)$
$(5\ 10\ 20\ 17\ 11\ 22\ 21\ 19\ 15\ 7\ 14)$

$\gamma = (0\infty)(1\ 22)(2\ 11)(3\ 15)(4\ 17)(5\ 9)(6\ 19)$
$(7\ 13)(8\ 20)(10\ 16)(12\ 21)(14\ 18)$

FIG. 3.9. Generators of $PSL_2(23)$.

Conway then showed that M_{24} is the group generated by $PSL_2(23)$, together with the permutation

$$\delta:\quad x \mapsto \begin{cases} 9x^3 & x \notin Q \\ x^3/9 & x \in Q \end{cases}$$

where Q is the set of quadratic residues modulo 23 together with 0. Thinking of $9\infty^3$ as ∞, we set $(\infty)\delta = \infty$. The cycle structure of δ is

$$(\infty)(0)(3)(15)(1\ \ 18\ \ 4\ \ 2\ \ 6)(5\ \ 21\ \ 20\ \ 10\ \ 7)$$

$$(8\ \ 16\ \ 13\ \ 9\ \ 12)(11\ \ 19\ \ 22\ \ 14\ \ 17).$$

Note that $(Q)\delta = Q$.

We show here that the group M, generated by α, β, γ and δ, is quintuply transitive on the set Ω. The fact that $M = M_{24}$ appears in Appendix 4. The proof follows that of Conway [**25**, pp. 223, 224]. All of this material was assumed in his original paper [**24**].

Conway used the term "shape" to describe the cycle structure of a permutation. For example, α has "shape" $1^1 23^1$ while the "shapes" of β, γ and δ are, respectively, $1^2 11^2$, 2^{12}, and $1^4 5^4$. Instead of his term "shape," we shall use the term *structure* (since "shape" was used already in referring to the Leech lattice Λ).

Since

$$\begin{aligned}\delta\alpha^2 = &(\infty)(4)(7)(0\ 2\ 8\ 18\ 6\ 3\ 5)(1\ 20\ 12\ 10\ 9\ 14\ 19)\\ &(11\ 21\ 22\ 16\ 15\ 17\ 13)\end{aligned}$$

$$\begin{aligned}(\alpha\delta)^3 = &(\infty)(0)(1)(5)(6)(18)(20)(22)(2\ 3)(4\ 15)(7\ 8)\\ &(9\ 11)(10\ 19)(12\ 16)(13\ 21)(14\ 17)\end{aligned}$$

and

$$\begin{aligned}(\alpha^{13}\gamma\delta^2)^3 = &(\infty\ 1\ 5\ 12)(0\ 2\ 11\ 22)(3\ 17\ 10\ 21)\\ &(4\ 7\ 8\ 19)(6\ 13\ 16\ 20)(9\ 18\ 15\ 14),\end{aligned}$$

M also has elements with structures $1^3 7^3$, $1^8 2^8$ and 4^6. The presence of permutations with structures $1^1 23^1$ and 2^{12} already show that M is transitive. To see this, suppose that i and j are in Ω. If i and j are in the cycle of length twenty-three, then the application of the permutation with structure $1^1 23^1$ to the appropriate power sends i to j. If i is in the 1-cycle, then the application of the permutation with structure 2^{12} to i, sends i to some element in the 23-cycle. This, followed by the application of the permutation with structure $1^1 23^1$ to the appropriate power, sends i to j. For example, suppose ∞ is to be sent to 15. Then

$$(\infty)\gamma\alpha^{15} = (0)\alpha^{15} = 15.$$

If j is in the 1-cycle, the inverse of the preceding mapping sends i to j.

The permutation $\gamma\alpha^m\gamma$ has structure $1^1 23^1$ and if we pick the appropriate m, the 1-cycle can be made to contain any particular element of Ω we choose. In fact, it is easy to show [**106**, pp. 139, 140] that the entries in the cycles of $\gamma\alpha^m\gamma = \gamma^{-1}\alpha^m\gamma$ are those of α^m replaced by their images under γ. This means that the stabilizer of any point has a permutation with structure $1^1 23^1$ and, hence, is transitive on the rest of Ω.

Suppose next that $i \neq k$, $j \neq l$ are elements of Ω. We wish to show that there is a permutation in M which sends i to j and k to l. Since M *is* transitive, there is a permutation η which sends i to j and k to some m. If $m \neq l$, there is a permutation ν in the stabilizer of j which sends m to k. Then

$$(i)\eta\nu = (j)\nu = j$$

and

$$(k)\eta\nu = (m)\nu = l.$$

This shows that M is 2-ply transitive. By using permutations with structure $1^2 11^2$ and $1^3 7^3$, it can be shown that the stabilizer of two points of Ω is transitive on the remaining twenty-two points of Ω. By using permutations with structure $1^3 7^3$ and $1^4 5^4$, one can show that the stabilizer of any three points of Ω is transitive on the remaining twenty-one points. All told, M is quadruply transitive on Ω.

The subgroup of M fixing any set of four points as a whole (not their pointwise stabilizer) has permutations with structure $1^4 5^4$ and 4^6 and so is transitive on the remaining twenty points. This can be argued as follows. Quadruple transitivity allows the permutation with structure $1^4 5^4$ to fix the given 4-set pointwise. Similarly, the permutation with structure 4^6 has the 4-set as a 4-cycle. But then the respective remaining 5-cycles and 4-cycles overlap to yield the desired transitivity. This means that given two 5-sets of Ω, there is an element in

M which sends one 5-set to the other 5-set but does not necessarily send the ordered elements of one 5-set to a particular ordering of the elements of the other. However, the subgroup fixing any particular 5-set has permutations with structure $1^4 5^4$ and $1^8 2^8$ which induce permutations with structures 5^1 and $1^3 2$ on that 5-set. It is an easy and standard exercise to show that a permutation with structure 5^1 and one with structure $1^3 2$ generate S_5. This, together with the preceding calculations, shows that M is quintuply transitive. The rest of the proof that M equals M_{24} is found in Appendix 4.

Further information on the Mathieu groups can be found in [**21**] and in the references in [**85**].

D. The Group *N*. In this section we construct a group, called N by Conway, which will turn out to be a maximal subgroup of $\cdot 0$ and will be of aid in determining the order of $\cdot 0$.

Recall that $\cdot 0$ is the group of symmetries of P and that

$$\mathbf{v}_i = (0, 0, \ldots, \underbrace{1}_{i\text{th coordinate}}, \ldots, 0)$$

where the twenty-four coordinate positions are indexed by $\Omega = \{0, 1, 2, \ldots, 22, \infty\}$.

Any permutation π of Ω that takes $R(C)$ into itself can be extended to a symmetry of P defined by $(\mathbf{v}_i)\pi = \mathbf{v}_{(i)\pi}$. Because the 8-sets generate $R(C)$, the set of all such permutations π is the automorphism group of a Steiner system $S(5, 8, 24)$ so that the set of all such symmetries forms a subgroup M of $\cdot 0$ isomorphic to M_{24}. Thus M_{24} is a subgroup of $\cdot 0$.

Any subset S of Ω determines an isometry, ϵ_S, of E^{24}, namely the reflection in a hyperplane of dimension $|S|$ defined by $(\mathbf{v}_i)\epsilon_S = -\mathbf{v}_i$ if i is in S and $(\mathbf{v}_i)\epsilon_S = \mathbf{v}_i$ if i is not in S. The map ϵ_S is in $\cdot 0$ provided S is in $R(C)$. Here "S in $R(C)$" means that the elements of S correspond exactly to coordinate positions of the 1's of some vector in $R(C)$. The group of all such ϵ_S's is abelian (which the reader can easily check), isomorphic to the additive group of $R(C)$, and will be designated by $E = E_{12}$ (not to be confounded with Euclidean 12-space) since $R(C)$ has 2^{12} vectors.

Conway defined $N = N_{24}$ to be the subgroup $EM = \{\epsilon\pi \mid \epsilon \in E, \pi \in M\}$ of $\cdot 0$. That N is indeed a subgroup of $\cdot 0$ follows from the fact that for S in $R(C)$ and π in M

$$\pi\epsilon_S = \epsilon_{(S)\pi^{-1}}\pi, \qquad (3.10)$$

so that EM is closed. In fact, N is a *semidirect product* of E by M. Recall that a group G is a semidirect product of K by H, K and H subgroups of G, in case

1. K is normal in G,
2. $KH = G$,
3. $K \cap H = \{1\}$. [**106**, p. 134]

Conditions (2) and (3) are equivalent to the assertion that each g in G is uniquely expandable in the form kh, k in K, h in H. Conway noted that N is a *splitting extension* of E by M which is equivalent to being a semidirect product of E by M [**24**, p. 81], [**106**, p. 134]. The three criteria are easy to establish. (The first follows immediately from Equation 3.10.)

To Conway, the subgroup N of $\cdot 0$ was the set of "visible" symmetries, generated by reflections in certain coordinate planes and by permutations of coordinates which were also symmetries of P [**28**]. These he called "fairly obvious," and were part of his first twelve-hour discovery.

Before establishing a couple of facts about the subgroup N of $\cdot 0$, we wish to sketch briefly how Conway used N to establish the order of $\cdot 0$.

He first assumed that N is not all of $\cdot 0$:

> I supposed that the visible group [N] wasn't all. I took as a small axiom that there was another symmetry that I didn't know. No statement about anything else. Then I showed that in that case

> there would have to be a symmetry [outside N] taking any vertex of P to any other vertex of P. . . . [26], [28]

This means that if H, a subgroup of $\cdot 0$, strictly contains N, H is transitive on the set of vertices, Λ_2, of P. Let $\mathbf{x}$ be a member of Λ_2 and $H_{\mathbf{x}}$, the elements of H fixing $\mathbf{x}$ (the *stabilizer* of $\mathbf{x}$ with respect to H). Using the fact that $|\overline{\mathbf{x}}|$, the size of the orbit of $\mathbf{x}$ under H, equals the index of $H_{\mathbf{x}}$ in H, $[H:H_{\mathbf{x}}]$, Conway claimed that $|H| = 196560\,|H_{\mathbf{x}}|$. He further showed that $|H_{\mathbf{x}}| = 93150\,|H_{\mathbf{x},\mathbf{y}}|$ where $H_{\mathbf{x},\mathbf{y}}$ is $(H_{\mathbf{x}})\mathbf{y}$, the stabilizer of $\mathbf{y}$ with respect to $H_{\mathbf{x}}$, of two orthogonal vertices (vertices $\mathbf{x}$, $\mathbf{y}$ such that their *inner product* $(\mathbf{x}, \mathbf{y})$ is zero) in Λ_2. However, at that point, he was able to show that $H_{\mathbf{x},\mathbf{y}} = N_{\mathbf{x},\mathbf{y}}$, a group of order $2^{10}|M_{22}|$ which he denoted N_{22}. This means that the order of H, a subgroup of $\cdot 0$ strictly containing N, was completely determined. Consequently, H must be all of $\cdot 0$ and N is maximal in $\cdot 0$. The only tasks left were to prove the above assertions. We now show how he accomplished this task.

THEOREM 3.1. *Any symmetry λ of $\cdot 0$ which fixes a coordinate vector $\mathbf{v}_i$ is in N.*

Assume that $(\mathbf{v}_i)\lambda = \mathbf{v}_i$, λ in $\cdot 0$. Let $\mathbf{v}_j$ be another coordinate vector and let $\mathbf{w}_j = (\mathbf{v}_j)\lambda$. Since λ preserves angle, $\mathbf{w}_j$ is orthogonal to $\mathbf{v}_i$. Furthermore, since $4\mathbf{v}_i + 4\mathbf{v}_j$ is in Λ_2, so is $4\mathbf{v}_i + 4\mathbf{w}_j$. The enumeration of Λ_2 in Table 3.7 shows that $\mathbf{w}_j = \pm\mathbf{v}_k$ for some k in Ω since $8\mathbf{w}_j$ is in Λ, has length eight, and $4\mathbf{w}_j$ is not in Λ. Distinct values of j must give distinct values of the corresponding k. Then $\lambda = \epsilon_S\pi$ for some permutation π and some subset S in Ω. The nonzero coordinates of $(2\mathbf{v}_K)\lambda$, K an 8-set in $R(C)$, are in the coordinate positions of $(K)\pi$. Thus $(K)\pi$ is also an 8-set in $R(C)$. Consequently, π preserves the 8-sets as a whole, and we conclude that π is in M. Since $\lambda = \epsilon_S\pi$, λ sends $\mathbf{v}_\Omega - 4\mathbf{v}_\infty$ to a lattice point of the same shape. First, note that the coordinates of $\mathbf{v}_\Omega - 4\mathbf{v}_\infty$ are congruent to 1 modulo 4. Then, because ϵ_S sends $\mathbf{v}_i$ to

$-\mathbf{v}_i$ if i is in S, the coordinates of $(\mathbf{v}_\Omega - 4\mathbf{v}_\infty)\lambda$ congruent to 3 modulo 4 are in the places of $(S)\pi$. Consequently, from the definition of Λ, $(S)\pi$, and thus S, must be in $R(C)$. This shows that $\lambda = \epsilon_S \pi$ is in N and proves the theorem. ■

A second fact that we will need is expressed in the next theorem.

THEOREM 3.2. *Any symmetry* λ *of* $\cdot 0$ *which takes* Λ_2^4 *into itself is in N.*

Recall first that Λ_2^4 is the set of vertices of P which have shape $(0^{22}, \pm 4^2)$. We will have recourse to the following lemmas.

LEMMA 3.3. *No symmetry in* $\cdot 0$ *has either prime order* $p > 23$, *or order* $13 \cdot 23$.

The proof of this lemma is quite algebraic. Let λ be a symmetry in $\cdot 0$ other than the identity. Suppose A is the matrix of λ and that $A^p = I$ for some prime p. Let $T(x)$ be the minimal polynomial for A. (We assume that $T(x)$ is monic.) Then $T(x)$ divides the polynomial $x^p - 1$ since A satisfies the equation $x^p - 1 = 0$. Furthermore, $T(x)$ is in $Q[x]$, since A has rational entries. Since

$$x^p - 1 = (x-1)(x^{p-1} + x^{p-2} + \cdots + x + 1),$$

the second factor of which is irreducible over $Q[x]$, and since $T(x) \neq x - 1$, $T(x)$ divides $x^{p-1} + x^{p-2} + \cdots + x + 1$ and so is equal to it. But A also satisfies its characteristic equation; thus

$$T(x) = (x^{p-1} + x^{p-2} + \cdots + x + 1)$$

is a factor of the characteristic polynomial of A, which has degree 24. Consequently, $p - 1 \leq 24$. Moreover, since p is prime, $p \leq 23$.

A similar argument, which proves the second part of the lemma, appears in Appendix 5. ■

LEMMA 3.4. *Let G be a group which acts on a set X. Let G_x be the stabilizer of a point x of X, and let $\bar{x} = \{(x)g \mid g \in G\}$ be the orbit of x. Then $[G:G_x] = |\bar{x}|$.*

For g_1, g_2 in G, $(x)g_1 = (x)g_2 \Leftrightarrow (x)g_1g_2^{-1} = x \Leftrightarrow g_1g_2^{-1}$ is in $G_x \Leftrightarrow g_1G_x = g_2G_x$ as (left) cosets. This implies that the map $gG_x \mapsto (x)g$ is a bijection from the family of the cosets of G_x in G onto $\bar{x}$. In particular, if $|\bar{x}|$ is finite, then $[G:G_x] = |\bar{x}|$. ∎

With these preliminaries disposed of, we now prove Theorem 3.2.

Let H be the set of all symmetries fixing Λ_2^4 as a whole. We wish to show that H is a subset of N. Let $N_{\mathbf{x}}$ denote the subgroup of N fixing $\mathbf{x}$ and $\Lambda_2^4(\mathbf{x})$ be the set of all vectors in Λ_2^4 orthogonal to $\mathbf{x}$, where $\mathbf{x} = 4\mathbf{v}_i + 4\mathbf{v}_j$. Since every element of N only permutes coordinates or changes the signs of the coordinates of a vector, N maps Λ_2^4 into itself. Furthermore, since $\cdot 0$ preserves angle, $N_{\mathbf{x}}$ fixes $\Lambda_2^4(\mathbf{x})$ as a whole.

An *orbit* of $\Lambda_2^4(\mathbf{x})$ under $N_{\mathbf{x}}$ consists of an element $\mathbf{y}$ of $\Lambda_2^4(\mathbf{x})$, together with all elements of the form $(\mathbf{y})\lambda$ where λ ranges over $N_{\mathbf{x}}$. Any two distinct such orbits are disjoint, and for any two elements $\mathbf{a}$ and $\mathbf{b}$ in one orbit, there is some isometry in $N_{\mathbf{x}}$ which maps $\mathbf{a}$ to $\mathbf{b}$.

It is easy to see that the $2^2 \cdot \binom{22}{2} = 924$ vectors of the form $\pm 4\mathbf{v}_h \pm 4\mathbf{v}_k$ are orthogonal to $4\mathbf{v}_i + 4\mathbf{v}_j$, h, i, j, k distinct. So are the two vectors $\pm(4\mathbf{v}_i - 4\mathbf{v}_j)$. This set of 926 vectors forms $\Lambda_2^4(\mathbf{x})$. Since M is quintuply transitive, it is doubly transitive and there is an element of $N_{\mathbf{x}}$ which sends $4\mathbf{v}_i - 4\mathbf{v}_j$ to its negative. Moreover, no element of $N_{\mathbf{x}}$ can send $4\mathbf{v}_i - 4\mathbf{v}_j$ to one of the other 924 vectors in $\Lambda_2^4(\mathbf{x})$. This means that the two points $4\mathbf{v}_i - 4\mathbf{v}_j$ and $-(4\mathbf{v}_i - 4\mathbf{v}_j)$ form one orbit. Similarly, the other 924 vectors in $\Lambda_2^4(\mathbf{x})$ form another orbit.

From the inclusion $N \subset H$, it follows that $N_{\mathbf{x}} \subset H_{\mathbf{x}}$. Thus, the orbits of $H_{\mathbf{x}}$ in $\Lambda_2^4(\mathbf{x})$ are unions of those of $N_{\mathbf{x}}$. This means that the orbits of $H_{\mathbf{x}}$ are either those of $N_{\mathbf{x}}$ or a single orbit consisting of all 926 vectors in $\Lambda_2^4(\mathbf{x})$. Suppose the latter case occurs. Let $\bar{\mathbf{y}} = \{(\mathbf{y})\lambda \mid \lambda \text{ in } H_{\mathbf{x}}\}$ be the orbit of $\mathbf{y}$ in $\Lambda_2^4(\mathbf{x})$. By Lemma 3.4, $|\bar{\mathbf{y}}|$ and thus $[H_{\mathbf{x}}:H_{\mathbf{x},\mathbf{y}}] = 926 = 2\cdot 463$, where $[H_{\mathbf{x}}:H_{\mathbf{x},\mathbf{y}}]$ is the index of $H_{\mathbf{x},\mathbf{y}}$ in $H_{\mathbf{x}}$. By Cauchy's Theorem, $H_{\mathbf{x}}$ and hence $\cdot 0$, has an element of order 463, in contradiction with Lemma 3.3.

As a result, $H_{\mathbf{x}}$ has two orbits on $\Lambda_2^4(\mathbf{x})$ and must map $4\mathbf{v}_i - 4\mathbf{v}_j$ either to itself or its negative. In the first case, λ would send $\mathbf{v}_i$ to $\mathbf{v}_i$ and hence be in N by Theorem 3.1. In the second case, λ would send $\mathbf{v}_i$ to $\mathbf{v}_j$. As in the proof of Theorem 3.1, λ would then send $4\mathbf{v}_i + 4\mathbf{v}_h$, $h \neq j$, to a vector of the form $4\mathbf{v}_j \pm 4\mathbf{v}_k$, implying that λ is in N. Hence, $H_{\mathbf{x}} \subset N$. Thus $H_{\mathbf{x}} \subset N_{\mathbf{x}}$. This shows that $H \subset N$ and establishes Theorem 3.2. ■

Before leaving the study of N to turn to the calculation of the order of $\cdot 0$, we examine the generators of N and show that N is a proper subgroup of $\cdot 0$.

We have already shown that M_{24} is generated by the four permutations:

$$\alpha = (\infty)(0\ 1\ 2\ 3\ \ldots\ 22)$$

$$\beta = (\infty)(0)(1\ 2\ 4\ 8\ 16\ 9\ 18\ 13\ 3\ 6\ 12)\\(5\ 10\ 20\ 17\ 11\ 22\ 21\ 19\ 15\ 7\ 14)$$

$$\gamma = (0\ \infty)(1\ 22)(2\ 11)(3\ 15)(4\ 17)(5\ 9)(6\ 19)\\(7\ 13)(8\ 20)(10\ 16)(12\ 21)(14\ 18)$$

and

$$\delta = (\infty)(0)(3)(15)(1\ 18\ 4\ 2\ 6)(5\ 21\ 20\ 10\ 7)\\(8\ 16\ 13\ 9\ 12)(11\ 19\ 22\ 14\ 17)$$

—see Figure 3.9 ff. Furthermore, it was noted that β is superfluous. These permutations in M_{24} correspond to the following symmetries in M, the isomorphic copy of M_{24} in $\cdot 0$:

$$\alpha: \quad \mathbf{v}_i \mapsto \mathbf{v}_{i+1}$$

$$\beta: \quad \mathbf{v}_i \mapsto \mathbf{v}_{2i}$$

$$\gamma: \quad \mathbf{v}_i \mapsto \mathbf{v}_{-1/i}$$

and

$$\delta: \quad \mathbf{v}_i \mapsto \begin{cases} \mathbf{v}_{9i^3} & i \text{ not in } Q \\ \mathbf{v}_{i^3/9} & i \text{ in } Q, \end{cases}$$
$$\mathbf{v}_\infty \mapsto \mathbf{v}_\infty,$$

where Q is the set of quadratic residues modulo 23 together with 0.

The context will make it clear whether α, β, etc., denotes a permutation in M_{24} or the corresponding symmetry in M. Note that α, β, γ, together with δ, generate M.

Define $\epsilon = \epsilon_Q$ by

$$\epsilon: \quad \mathbf{v}_i \mapsto \begin{cases} \mathbf{v}_i & i \text{ not in } Q \\ -\mathbf{v}_i & i \text{ in } Q. \end{cases}$$

We now claim that α, β, γ, δ, together with ϵ, generate N.

LEMMA 3.5. *The symmetries α, β, γ, δ and ϵ generate N.*

With appropriate interchange of indices, Q is a 12-set in $R(C)$. In the sequel, we shall always refer to 8-sets, etc., in $R(C)$, assuming that $R(C)$ has the *appropriate column interchanges.* Since $M = M_{24}$ preserves $R(C)$, 8-sets are sent to 8-sets, 12-sets are sent to 12-sets, etc. It will now be shown that α, δ, and ϵ generate a reflection ϵ_K, K an 8-set in $R(C)$, where

$$\epsilon_K: \quad \mathbf{v}_i \mapsto \begin{cases} \mathbf{v}_i & i \text{ not in } K \\ -\mathbf{v}_i & i \text{ in } K. \end{cases}$$

Recall that $Q = \{0, 1, 2, 3, 4, 6, 8, 9, 12, 13, 16, 18\}$.

The elements of Ω will be listed in a column and the action of α, for example, on $\mathbf{v}_i$ will be written

$$i \overset{\alpha}{\mapsto} i + 1.$$

Similarly, the action of ϵ on $\mathbf{v}_i$, i in Q, will be written

$$i \overset{\epsilon}{\mapsto} -i$$

instead of

$$\mathbf{v}_i \overset{\epsilon}{\mapsto} -\mathbf{v}_i.$$

Notice in particular the effect of the mapping

$$\epsilon\alpha\delta\alpha\epsilon\alpha^{-1}\delta^{-1}\alpha^{-1}$$

on the 12-set Q in Figure 3.11. This is the same as that of ϵ_K where

$$K = \{0, 1, 4, 5, 11, 12, 14, 22\}.$$

∞	$\xrightarrow{\epsilon} \infty$	$\xrightarrow{\alpha} \infty$	$\xrightarrow{\delta} \infty$	$\xrightarrow{\alpha} \infty$	$\xrightarrow{\epsilon} \infty$	$\xrightarrow{\alpha^{-1}\delta^{-1}\alpha^{-1}} \infty$
0	−0	−1	−18	−19	−19	−0
1	−1	−2	−6	−7	−7	−1
2	−2	−3	−3	−4	4	2
3	−3	−4	−2	−3	3	3
4	−4	−5	−21	−22	−22	−4
5	5	6	1	2	−2	−5
6	−6	−7	−5	−6	6	6
7	7	8	16	17	17	7
8	−8	−9	−12	−13	13	8
9	−9	−10	−7	−8	8	9
10	10	11	19	20	20	10
11	11	12	8	9	−9	−11
12	−12	−13	−9	−10	−10	−12
13	−13	−14	−17	−18	18	13
14	14	15	15	16	−16	−14
15	15	16	13	14	14	15
16	−16	−17	−11	−12	12	16
17	17	18	4	5	5	17
18	−18	−19	−22	0	0	18
19	19	20	10	11	11	19
20	20	21	20	21	21	20
21	21	22	14	15	15	21
22	22	0	0	1	−1	−22

FIG. 3.11. The generating of ϵ_K.

Let L be any other 8-set in $R(C)$ and let θ in M map K to L. (Such a θ must exist since the quintuple transitivity of M, together with the facts that any 5-set of Ω determines a unique 8-set in $R(C)$ and that M preserves $R(C)$ as a whole, implies that M is transitive on the 8-sets.) Then $\epsilon_L = \theta^{-1}\epsilon_K\theta$ can be generated by ϵ, α, and δ, together with the appropriate θ from M. By Corollary A2.7, any ϵ_D can be generated for D in $R(C)$. Thus E_{12} and hence all of N is generated by α, β, γ, δ and ϵ, and the lemma is proved. ∎

The matrix representations of these five generators of N are given in Figures 3.12 through 3.16. (Keep in mind that we are now using algebraic notation—not functional notation. This means that all vectors will be multiplied *on the right* by these matrices.) Each row (and column) has exactly one ± 1 and all the rest 0's. The rows represent the images of the basis vectors, in the order listed at the beginning of this section, on the group N.

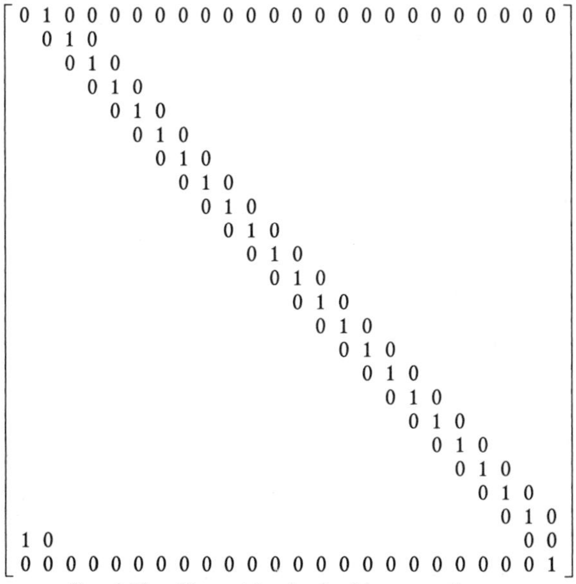

FIG. 3.12. The matrix of order 24 representing α.

$$\begin{bmatrix}
1 & 0 & 0 & 0 & 0 & 0 & 0 & 0 & 0 & 0 & 0 & 0 & 0 & 0 & 0 & 0 & 0 & 0 & 0 & 0 & 0 & 0 & 0 & 0 \\
& 0 & 1 & 0 \\
&&& 0 & 1 & 0 \\
&&&&& 0 & 1 & 0 \\
&&&&&&& 0 & 1 & 0 \\
&&&&&&&&& 0 & 1 & 0 \\
&&&&&&&&&&& 0 & 1 & 0 \\
&&&&&&&&&&&&& 0 & 1 & 0 \\
&&&&&&&&&&&&&&& 0 & 1 & 0 \\
&&&&&&&&&&&&&&&&& 0 & 1 & 0 \\
&&&&&&&&&&&&&&&&&&& 0 & 1 & 0 \\
&&&&&&&&&&&&&&&&&&&&& 0 & 1 & 0 \\
0 & 1 & 0 & 0 & 0 & 0 & 0 & 0 & 0 & 0 & 0 & 0 & 0 & 0 & 0 & 0 & 0 & 0 & 0 & 0 & 0 & 0 & 0 & 0 \\
&& 0 & 1 & 0 \\
&&&& 0 & 1 & 0 \\
&&&&&& 0 & 1 & 0 \\
&&&&&&&& 0 & 1 & 0 \\
&&&&&&&&&& 0 & 1 & 0 \\
&&&&&&&&&&&& 0 & 1 & 0 \\
&&&&&&&&&&&&&& 0 & 1 & 0 \\
&&&&&&&&&&&&&&&& 0 & 1 & 0 \\
&&&&&&&&&&&&&&&&&& 0 & 1 & 0 \\
&&&&&&&&&&&&&&&&&&&& 0 & 1 & 0 \\
0 & 0 & 0 & 0 & 0 & 0 & 0 & 0 & 0 & 0 & 0 & 0 & 0 & 0 & 0 & 0 & 0 & 0 & 0 & 0 & 0 & 0 & 0 & 1
\end{bmatrix}$$

FIG. 3.13. The matrix of order 24 representing β.

$$\begin{bmatrix}
0 & 0 & 0 & 0 & 0 & 0 & 0 & 0 & 0 & 0 & 0 & 0 & 0 & 0 & 0 & 0 & 0 & 0 & 0 & 0 & 0 & 0 & 0 & 1 \\
&&&&&&&&&&&&&&&&&&&&&& 1 & 0 \\
0 & 0 & 0 & 0 & 0 & 0 & 0 & 0 & 0 & 0 & 0 & 1 \\
&&&&&&&&&&&&&&& 1 & 0 & 0 & 0 & 0 & 0 & 0 & 0 & 0 \\
&&&&&&&&&&&&&&&&& 1 & 0 & 0 & 0 & 0 & 0 & 0 \\
0 & 0 & 0 & 0 & 0 & 0 & 0 & 0 & 0 & 1 \\
&&&&&&&&&&&&&&&&&&& 1 & 0 & 0 & 0 & 0 \\
&&&&&&&&&&&&& 1 & 0 & 0 & 0 & 0 & 0 & 0 & 0 & 0 & 0 & 0 \\
&&&&&&&&&&&&&&&&&&&& 1 & 0 & 0 & 0 \\
0 & 0 & 0 & 0 & 0 & 1 \\
&&&&&&&&&&&&&&&& 1 & 0 & 0 & 0 & 0 & 0 & 0 & 0 \\
0 & 0 & 1 \\
&&&&&&&&&&&&&&&&&&&&& 1 & 0 & 0 \\
0 & 0 & 0 & 0 & 0 & 0 & 0 & 1 \\
&&&&&&&&&&&&&&&&&& 1 & 0 & 0 & 0 & 0 & 0 \\
0 & 0 & 0 & 1 \\
0 & 0 & 0 & 0 & 0 & 0 & 0 & 0 & 0 & 0 & 1 \\
0 & 0 & 0 & 0 & 1 \\
&&&&&&&&&&&&&& 1 & 0 & 0 & 0 & 0 & 0 & 0 & 0 & 0 & 0 \\
0 & 0 & 0 & 0 & 0 & 0 & 1 \\
0 & 0 & 0 & 0 & 0 & 0 & 0 & 0 & 1 \\
&&&&&&&&&&&& 1 & 0 & 0 & 0 & 0 & 0 & 0 & 0 & 0 & 0 & 0 & 0 \\
0 & 1 \\
1 & 0 & 0 & 0 & 0 & 0 & 0 & 0 & 0 & 0 & 0 & 0 & 0 & 0 & 0 & 0 & 0 & 0 & 0 & 0 & 0 & 0 & 0 & 0
\end{bmatrix}$$

FIG. 3.14. The matrix of order 24 representing γ.

$$
\left[\begin{array}{cccccccccccccccccccccccc}
1&0&0&0&0&0&0&0&0&0&0&0&0&0&0&0&0&0&0&0&0&0&0&0\\
&&&&&&&&&&&&&&&&&&1&0&0&0&0&0\\
0&0&0&0&0&0&1&&&&&&&&&&&&&&&&&\\
0&0&0&1&&&&&&&&&&&&&&&&&&&&\\
0&0&1&&&&&&&&&&&&&&&&&&&&&\\
&&&&&&&&&&&&&&&&&&&&&1&0&0\\
0&1&&&&&&&&&&&&&&&&&&&&&&\\
0&0&0&0&0&1&&&&&&&&&&&&&&&&&&\\
&&&&&&&&&&&&&&&&1&0&0&0&0&0&0&0\\
&&&&&&&&&&&&1&0&0&0&0&0&0&0&0&0&0&0\\
0&0&0&0&0&0&0&1&&&&&&&&&&&&&&&&\\
&&&&&&&&&&&&&&&&&&&1&0&0&0&0\\
0&0&0&0&0&0&0&0&1&&&&&&&&&&&&&&&\\
0&0&0&0&0&0&0&0&0&1&&&&&&&&&&&&&&\\
&&&&&&&&&&&&&&&&&1&0&0&0&0&0&0\\
&&&&&&&&&&&&&&&1&0&0&0&0&0&0&0&0\\
&&&&&&&&&&&&&1&0&0&0&0&0&0&0&0&0&0\\
0&0&0&0&0&0&0&0&0&0&0&1&&&&&&&&&&&&\\
0&0&0&0&1&&&&&&&&&&&&&&&&&&&\\
&&&&&&&&&&&&&&&&&&&&&&1&0\\
0&0&0&0&0&0&0&0&0&0&1&&&&&&&&&&&&&\\
&&&&&&&&&&&&&&&&&&&&1&0&0&0\\
&&&&&&&&&&&&&&1&0&0&0&0&0&0&0&0&0\\
0&0&0&0&0&0&0&0&0&0&0&0&0&0&0&0&0&0&0&0&0&0&0&1
\end{array}\right]
$$

FIG. 3.15. The matrix of order 24 representing δ.

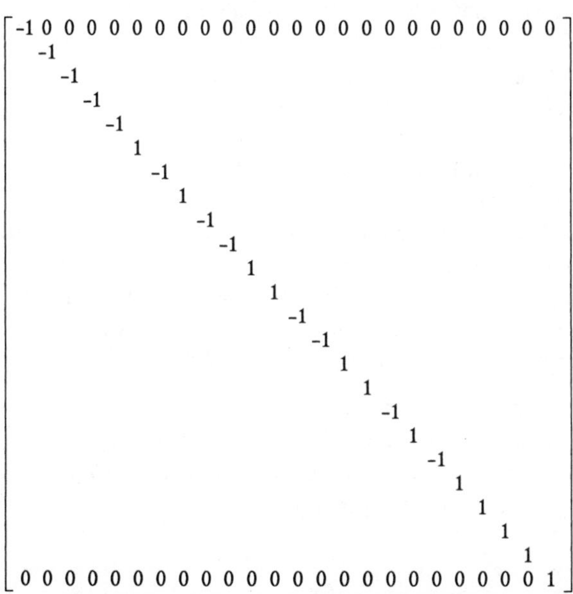

FIG. 3.16. The matrix of order 24 representing ϵ.

Since each row of the matrix representations of α, β, γ, δ and ϵ has a single ± 1 and the rest of the entries 0, these matrices fix each subset Λ_2^2, Λ_2^3 and Λ_2^4 of Λ_2 as a whole. It was at this point that Conway suspected that $\cdot 0$ contained a symmetry outside of N, i.e., outside of those generated by α, β, γ, δ and ϵ. This was his "small axiom that there was another symmetry...." The additional symmetry which appears in his paper [**24**] is not the first one Conway discovered. It was found "a considerable time later" and "just makes everything sort of very neat looking" [**27**]. The original one "was really quite a nice-looking matrix, but involves a lot more work" [**27**]. We now show how Conway constructed the additional symmetry which he gave in [**24**].

THEOREM 3.6. *N is a proper subgroup of $\cdot 0$.*

Let $T = T_0$ be any 4-set of Ω. As we have already shown (via Corollary A2.5), T lies in five 8-sets given by $T + T_1$, $T + T_2$, $\ldots$, $T + T_5$, where T_i is the complement of T in the ith 8-set. This means that Ω is the disjoint union of the six 4-sets, $T = T_0, T_1, T_2, \ldots, T_5$. This family, B, of 4-sets, will be used to generate a symmetry ζ outside N but in $\cdot 0$.

Define $\eta = \eta_B$ to be the operation taking $\mathbf{v}_i$ to $\mathbf{v}_i - \frac{1}{2}\mathbf{v}_{T_j}$ whenever i is in T_j. Then define $\zeta_T = \eta\epsilon_T$. The mapping ζ_T will be shown to be in $\cdot 0$ by examining its effect on the 760 element spanning set for Λ given on page 125, i.e., the 759 vectors of the form $\mathbf{v}_K$ where K is an 8-set in $R(C)$ together with $\mathbf{v}_\Omega - 4\mathbf{v}_\infty$.

Suppose K is one of the 759 8-sets of $R(C)$. There are essentially three ways, not counting permutation of the indices, in which K can intersect the T_i:

1. K is the union of T_0 and T_1,
2. K intersects T_0, T_1, T_2 and T_3 each in two points,
3. K intersects T_0 in three points and each of $T_1, T_2, \ldots,$ T_5 in exactly one point.

This list is complete because if, for example, K intersects T_0 in two points and some other T_i, say T_1, in more than two points, K would have at least five elements in common with the 8-set $T_0 + T_1$. This is not possible since K is distinct from $T_0 + T_1$. (By Corollary A2.3, distinct 8-sets have either zero, two or four elements in common.) This also shows that if K and T_0 have exactly three elements in common, K intersects each of the other T_i in exactly one element, since each of $T_0 + T_1, T_0 + T_2, \ldots,$ is also an 8-set.

For each of the three cases, consider the vectors $\mathbf{y} = (2\mathbf{v}_K)\eta\epsilon_{T_0}$ and $\mathbf{z} = (2\mathbf{v}_K)\eta\epsilon_{T_5}$. For convenience of notation, assume $T_0 = \{0, 1, 2, 3\}$, $T_1 = \{4, 5, 6, 7\}$, $T_2 = \{8, 9, 10, 11\}$, etc.

In the first case we may assume $K = \{0, 1, 2, 3, \ldots, 7\}$. Thus

$$(\mathbf{v}_0)\eta = \mathbf{v}_0 - \tfrac{1}{2}(\mathbf{v}_0 + \mathbf{v}_1 + \mathbf{v}_2 + \mathbf{v}_3)$$

$$(\mathbf{v}_1)\eta = \mathbf{v}_1 - \tfrac{1}{2}(\mathbf{v}_0 + \mathbf{v}_1 + \mathbf{v}_2 + \mathbf{v}_3)$$

$$\vdots$$

and

$$(\mathbf{v}_7)\eta = \mathbf{v}_7 - \tfrac{1}{2}(\mathbf{v}_4 + \mathbf{v}_5 + \mathbf{v}_6 + \mathbf{v}_7).$$

Then

$$\begin{aligned}(\mathbf{v}_K)\eta &= \mathbf{v}_K - 2\mathbf{v}_K\\ &= -\mathbf{v}_K\end{aligned}$$

so that

$$\begin{aligned}\mathbf{y} &= (2\mathbf{v}_K)\eta\epsilon_{T_0}\\ &= (-2\mathbf{v}_K)\epsilon_{T_0}\\ &= 2\mathbf{v}_{T_0} - 2\mathbf{v}_{T_1}\end{aligned}$$

is in Λ_2 as is

$$\begin{aligned} \mathbf{z} &= (2\mathbf{v}_K)\eta\epsilon_{T_5} \\ &= (-2\mathbf{v}_K)\epsilon_{T_5} \\ &= -2\mathbf{v}_K. \end{aligned}$$

In the second case ($K = \{0, 1, 4, 5, 8, 9, 12, 13\}$, say),

$$\begin{aligned} \mathbf{y} &= (2\mathbf{v}_K)\eta\epsilon_{T_0} \\ &= (2(\mathbf{v}_K)\eta)\epsilon_{T_0} \\ &= (2(\mathbf{v}_K - \mathbf{v}_{T_0+T_1+T_2+T_3}))\epsilon_{T_0} \\ &= -(2(\mathbf{v}_{K+T_0+T_1+T_2+T_3}))\epsilon_{T_0} \\ &= 2\mathbf{v}_L - 2\mathbf{v}_M, \end{aligned}$$

where L is the 2-set in T_0 but not in K and M is the 6-set in $T_1 + T_2 + T_3$ but not in K. Again, $\mathbf{y}$ is in Λ_2, as is

$$\begin{aligned} \mathbf{z} &= (2\mathbf{v}_K)\eta\epsilon_{T_5} \\ &= -(2(\mathbf{v}_{K+T_0+T_1+T_2+T_3}))\epsilon_{T_5} \\ &= 2\mathbf{v}_{K+T_0+T_1+T_2+T_3} \end{aligned}$$

since $K + T_0 + T_1 + T_2 + T_3$ is an 8-set in $R(C)$.

The third case is a little more involved. (Let $K = \{0, 1, 2, 4, 8, 12, 16, 20\}$.)

$$\begin{aligned} \mathbf{y} &= (2\mathbf{v}_K)\eta\epsilon_{T_0} \\ &= (2(\mathbf{v}_K - \tfrac{1}{2}\mathbf{v}_{\Omega-T_0} - \tfrac{3}{2}\mathbf{v}_{T_0}))\epsilon_{T_0} \\ &= (2\mathbf{v}_K - \mathbf{v}_{\Omega-T_0} - 3\mathbf{v}_{T_0})\epsilon_{T_0} \\ &= 2\mathbf{v}_{K-T_0} - 2\mathbf{v}_{K\cap T_0} + 3\mathbf{v}_{T_0} - \mathbf{v}_{\Omega-T_0} \\ &= -\mathbf{v}_\Omega + 4\mathbf{v}_{T_0-K} + 2\mathbf{v}_K \\ &= (-\mathbf{v}_\Omega + 4\mathbf{v}_{T_0-K})\epsilon_K \end{aligned}$$

is in Λ_2. Similarly,

$$\begin{aligned}
\mathbf{z} &= (2\mathbf{v}_K)\eta\epsilon_{T_5} \\
&= (2\mathbf{v}_K - \mathbf{v}_{\Omega-T_0} - 3\mathbf{v}_{T_0})\epsilon_{T_5} \\
&= 2\mathbf{v}_{K-T_5} - 2\mathbf{v}_{K\cap T_5} - \mathbf{v}_{\Omega-(T_0+T_5)} + \mathbf{v}_{T_5} - 3\mathbf{v}_{T_0} \\
&= 2\mathbf{v}_{K-T_5} - 2\mathbf{v}_{K\cap T_5} - \mathbf{v}_{\Omega} + 2\mathbf{v}_{T_5} - 2\mathbf{v}_{T_0} \\
&= -\mathbf{v}_{\Omega} - 2\mathbf{v}_{T_0-K} + 2\mathbf{v}_{K-(T_0+T_5)} + 2\mathbf{v}_{T_5-K} \\
&= -\mathbf{v}_{\Omega} - 4\mathbf{v}_{T_0-K} + 2\mathbf{v}_{K+T_0+T_5} \\
&= (-\mathbf{v}_{\Omega} + 4\mathbf{v}_{T_0-K})\epsilon_{K+T_0+T_5}
\end{aligned}$$

which is in Λ_2 since $K + T_0 + T_5$ is an 8-set in $R(C)$.

The other spanning element is $\mathbf{v}_\Omega - 4\mathbf{v}_\infty$. If ∞ is in T, then

$$\begin{aligned}
(\mathbf{v}_\Omega - 4\mathbf{v}_\infty)\zeta_T &= (\mathbf{v}_\Omega - 4\mathbf{v}_\infty)\eta\epsilon_T \\
&= (-\mathbf{v}_\Omega - 4\mathbf{v}_\infty + 2\mathbf{v}_T)\epsilon_T \\
&= -\mathbf{v}_\Omega + 2\mathbf{v}_T + 4\mathbf{v}_\infty - 2\mathbf{v}_T \\
&= -(\mathbf{v}_\Omega - 4\mathbf{v}_\infty) \text{ is in } \Lambda_2.
\end{aligned}$$

In the same way, if ∞ is in $T_i \neq T$,

$$\begin{aligned}
(\mathbf{v}_\Omega - 4\mathbf{v}_\infty)\zeta_T &= (-\mathbf{v}_\Omega - 4\mathbf{v}_\infty + 2\mathbf{v}_{T_i})\epsilon_T \\
&= -\mathbf{v}_\Omega + 2\mathbf{v}_T - 4\mathbf{v}_\infty + 2\mathbf{v}_{T_i} \\
&= (-\mathbf{v}_\Omega + 4\mathbf{v}_\infty)\epsilon_{T+T_i}
\end{aligned}$$

is in Λ_2 since $T + T_i$ is an 8-set in $R(C)$. Thus ζ_T fixes Λ_2 as a whole.

An easy exercise now shows that $\zeta_T^2 = 1$, so that ζ_T is in $\cdot 0$ since it is invertible. Since any element of N fixed each Λ_2^i as a whole and since the third case showed that ζ_T sends an element of Λ_2^2 to one in Λ_2^3, ζ_T is outside N. This proves the theorem. ∎

Conway, in [**24**], had set $T = \{0, 3, 15, \infty\}$, so that the other 4-sets were:

$$\{1, 12, 21, 22\}, \{2, 7, 11, 13\}, \{4, 10, 16, 17\}$$

$$\{5, 6, 9, 19\} \text{ and } \{8, 14, 18, 20\}.$$

These are easy to check against Todd's list of 759 8-sets given in [**120**]. The matrix representation of ζ is given in Figure 3.17.

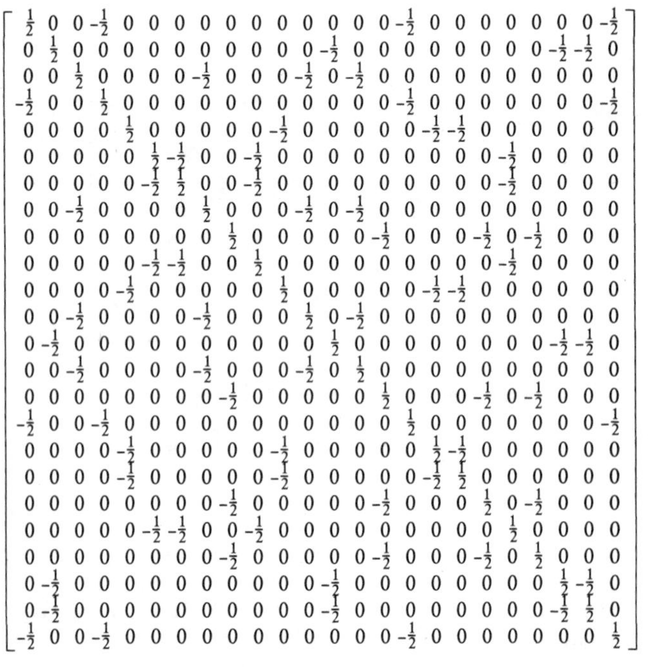

FIG. 3.17. The matrix of order 24 representing ζ.

E. The Order of $\cdot 0$. Let H be any subgroup of $\cdot 0$ strictly containing N. Conway's method of proof will show that, in fact, $H = \cdot 0$ and at the same time calculate the order of $\cdot 0$.

THEOREM 3.7. *Let H be any subgroup of $\cdot 0$ strictly containing N. Then H is transitive on Λ_2.*

This will be proved by showing that H has but a single orbit on Λ_2. The orbits of H on Λ_2 are unions of orbits of N on Λ_2 which, in turn, are the orbits of N on Λ_2^2, Λ_2^3 and Λ_2^4, since N fixes each of these sets as a whole. Consider first the set Λ_2^4. Since N is quintuply (and hence doubly transitive) on Ω, $4\mathbf{v}_1 + 4\mathbf{v}_2$ can be sent to $4\mathbf{v}_i + 4\mathbf{v}_j$ for any distinct i and j in Ω. If i is in K, and j is not in K for some K in $R(C)$, then ϵ_K, which is in N, sends $4\mathbf{v}_i + 4\mathbf{v}_j$ to $-4\mathbf{v}_i + 4\mathbf{v}_j$. Similarly, $4\mathbf{v}_i + 4\mathbf{v}_j$ can be sent to either $4\mathbf{v}_i - 4\mathbf{v}_j$ or $-4\mathbf{v}_i - 4\mathbf{v}_j$. Then N is transitive on Λ_2^4 and so has but a single orbit on Λ_2^4, namely, Λ_2^4. Similarly, Λ_2^2 and Λ_2^3 are single orbits of N.

The orbit of H containing Λ_2^4 cannot be just Λ_2^4 because any element of $\cdot 0$ that fixes Λ_2^4 as a whole is in N (Theorem 3.2). Suppose that the orbit is $\Lambda_2^4 \cup \Lambda_2^2$. Now

$$|\Lambda_2^4 \cup \Lambda_2^2| = 759\cdot 2^7 + \binom{24}{2}\cdot 2^2 = 2^4\cdot 3\cdot 23\cdot 89.$$

This means that for any $\mathbf{x}$ in the union,

$$[H:H_{\mathbf{x}}] = 2^4\cdot 3\cdot 23\cdot 89,$$

so that 89 divides $|\cdot 0|$ whence $\cdot 0$ has an element of order 89. This is impossible because Lemma 3.3 said that no element of $\cdot 0$ has prime order greater than 23. A similar argument shows that $\Lambda_2^4 \cup \Lambda_2^3$ is not an orbit of H so that, indeed, Λ_2 is a single orbit for H. ■

The next step is to establish some transitivity for $H_{\mathbf{x}}$.

THEOREM 3.8. *$H_{\mathbf{x}}$ is transitive on $\Lambda_2(\mathbf{x})$ for any $\mathbf{x}$ in Λ_2.*

First, let $\mathbf{x} = \mathbf{v}_\Omega - 4\mathbf{v}_\infty$. Since $(\mathbf{v}_\Omega - 4\mathbf{v}_\infty)\alpha = \mathbf{v}_\Omega - 4\mathbf{v}_\infty$, $N_{\mathbf{x}}$ contains the symmetry α of order 23, and so does $H_{\mathbf{x}}$. Since α moves each element in $\Lambda_2(\mathbf{x})$ (the set of elements orthogonal to $\mathbf{x}$ in Λ_2) the order of each orbit of $H_{\mathbf{x}}$ on $\Lambda_2(\mathbf{x})$ is a multiple of 23. This follows from the fact that sets of the form

$$A = \{\mathbf{y}, (\mathbf{y})\alpha, (\mathbf{y})\alpha^2, \ldots, (\mathbf{y})\alpha^{22}\}$$

are disjoint and fill out any such orbit.

We next show that 23 divides the order of each orbit of $H_{\mathbf{w}}$ on $\Lambda_2(\mathbf{w})$ for any $\mathbf{w}$ in Λ_2.

Since H is transitive on Λ_2, there is an element η in H such that $(\mathbf{x})\eta = \mathbf{w}$. Suppose $\mathbf{y}$ and $\mathbf{z}$ are in some orbit of $\Lambda_2(\mathbf{x})$ under $H_{\mathbf{x}}$. Then there is an element ζ in $H_{\mathbf{x}}$ which maps $\mathbf{y}$ to $\mathbf{z}$. Since $(\mathbf{x}, \mathbf{y}) = (\mathbf{x}, \mathbf{z}) = 0$,

$$\begin{aligned}(\mathbf{w}, (\mathbf{y})\eta) &= ((\mathbf{x})\eta, (\mathbf{y})\eta) = (\mathbf{x}, \mathbf{y}) = 0 = (\mathbf{x}, \mathbf{z}) \\ &= ((\mathbf{x})\eta, (\mathbf{z})\eta) = (\mathbf{w}, (\mathbf{z})\eta).\end{aligned}$$

Thus both $(\mathbf{y})\eta$ and $(\mathbf{z})\eta$ are in $\Lambda_2(\mathbf{w})$. But then

$$(\mathbf{w})\eta^{-1}\zeta\eta = (\mathbf{x})\zeta\eta = (\mathbf{x})\eta = \mathbf{w},$$

so that $\eta^{-1}\zeta\eta$ is in $H_{\mathbf{w}}$ and $(\mathbf{y})\eta\eta^{-1}\zeta\eta = (\mathbf{y})\zeta\eta = (\mathbf{z})\eta$. Thus $(\mathbf{y})\eta$ and $(\mathbf{z})\eta$ are in the same orbit of $\Lambda_2(\mathbf{w})$ under $H_{\mathbf{w}}$. It is now easy to show that the orbit containing $(\mathbf{y})\eta$ and $(\mathbf{z})\eta$ is at least as large as that containing $\mathbf{y}$ and $\mathbf{z}$ and that the converse is also true. This means that the orbits of $H_{\mathbf{w}}$ on $\Lambda_2(\mathbf{w})$ have the same size as those of $H_{\mathbf{x}}$ on $\Lambda_2(\mathbf{x})$ and thus are multiples of 23.

Now the orbits of $H_{\mathbf{w}}$ on $\Lambda_2(\mathbf{w})$ are unions of the orbits of $N_{\mathbf{w}}$ on $\Lambda_2(\mathbf{w})$. For convenience, the latter orbits will be calculated for $\mathbf{w} = 4\mathbf{v}_i + 4\mathbf{v}_j$. Two have already been done, namely the orbit consisting of the two points

$$\pm(4\mathbf{v}_i - 4\mathbf{v}_j)$$

which will be called 0_1, and the set 0_2 consisting of the $\binom{22}{2}\cdot 2^2$ points

$$\pm 4\mathbf{v}_h \pm 4\mathbf{v}_k,$$

h, i, j and k distinct. These are all the vectors in $\Lambda_2^4(\mathbf{w})$. There are two orbits of $N_\mathbf{w}$ on $\Lambda_2^2(\mathbf{w})$. To see this, we must first find $\Lambda_2^2(\mathbf{w})$. It is easy to see that any vector $2\mathbf{v}_K$ in Λ_2^2, where neither i nor j is in K, is orthogonal to $\mathbf{w}$. In fact, so are all vectors of the form $(2\mathbf{v}_K)\epsilon_D$ for such K and for D in $R(C)$. These vectors have shape $(0^{16}, \pm 2^8)$, with an even number of 2's and also of -2's, and the nonzero coordinates in the positions indicated by K, an 8-set in $R(C)$ not containing either i or j. For each set K, there are seven choices of sign and thus 2^7 possibilities. Of the 759 8-sets, $759 \cdot \frac{16}{24} = 506$ miss i and $506 \cdot \frac{15}{23} = 330$ also miss j. Denote by 0_3 the $330 \cdot 2^7 = 42240$ 8-sets of the form

$$(2\mathbf{v}_K)\epsilon_D,$$

K disjoint from $\{i, j\}$, and D in $R(C)$. The set $N_\mathbf{w}$ preserves this set as a whole, so that the orbits of $N_\mathbf{w}$ on 0_3 are subsets of 0_3. We now show that 0_3 is a single orbit for $N_\mathbf{w}$.

Let K be any 8-set disjoint from $\{i, j\}$, and T, a 4-set of K. Form the family $\{T, T_1, T_2, T_3, T_4, T_5\}$ of pairwise disjoint 4-sets whose union is Ω and any two of which form an 8-set. One, say T_1, is disjoint from $\{i, j\}$, whence ϵ_{T+T_1} is in $N_\mathbf{w}$. But then $(2\mathbf{v}_K)\epsilon_{T+T_1}$ has -2's in the positions of T, so that $N_\mathbf{w}$ moves $2\mathbf{v}_K$ to a vector with -2's in the positions of any four coordinates of $2\mathbf{v}_K$.

Suppose $K = \{a_1, a_2, \ldots, a_8\}$. From Table A2.3, 76 8-sets, other than K contain both a_1 and a_2, and 4 8-sets other than K contain a_1, a_2 and two other elements of K. Thus there are $77 - \binom{6}{2} \cdot 4 = 16$ 8-sets which intersect the 8-set K in exactly the elements a_1 and a_2. If one of these, say L, is disjoint from $\{i, j\}$, then ϵ_L is in $N_\mathbf{w}$ and $(2\mathbf{v}_K)\epsilon_L$ has -2's in the positions a_1 and a_2. We now show that at least one such L exists.

Assume that $K = T + T_1$ and $\{i, j\} \subset T_2$. Let L be an 8-set containing a_1, a_2, the two elements in $T_2 + \{i, j\}$, and one element from T_3. Then L is disjoint from $\{i, j\}$—otherwise L and $T + T_2$ would have five elements in common which is impossible. Similarly, one can pick such an L even when i and j are in distinct 4-sets.

It is not hard to show that ϵ_K is in $N_{\mathbf{w}}$ and that $(2\mathbf{v}_K)\epsilon_K = -2\mathbf{v}_K$. If an 8-set L intersects K in the two elements complement to a 6-set of K, and if L is disjoint from $\{i, j\}$, then $(2\mathbf{v}_K)\epsilon_L\epsilon_K$ has -2's in the positions of the 6-set. Thus 0_3 is a single orbit for the group $N_{\mathbf{w}}$ if it can be shown that $N_{\mathbf{w}}$ moves $2\mathbf{v}_K$ to any $2\mathbf{v}_L$ where L is also disjoint from $\{i, j\}$.

So far, we know that the orbits of $N_{\mathbf{w}}$ on 0_3 all have at least $2^7 = 128$ points. In fact, they have orders divisible by 128 because, for each vector $2\mathbf{v}_K$ in an orbit, there are 127 other vectors of the form $(2\mathbf{v}_K)\epsilon_D$, distinct from $2\mathbf{v}_K$, in the same orbit, where ϵ_D is in $N_{\mathbf{w}}$.

Let M be an 8-set which contains $\{i, j\}$ and any three points of K. Thus M must contain an additional point of K. The group M_{24} contains a permutation η which fixes i and j, and sends the three points of K to three points of L, say b_1, b_2 and b_3. Thus $(K)\eta$ must contain an additional point of L, say b_4, the image under η of the fourth point of $M \cap K$. If $(K)\eta = L$, we are done since the corresponding η in N is actually in $N_{\mathbf{w}}$. Otherwise, let b_5 be an additional element of $(K)\eta$, and b_6 an additional element of L (both outside $(M)\eta$). By Lemma A4.3, the subgroup of M_{24} fixing $(M)\eta$ elementwise contains a permutation ζ that sends b_5 to b_6. Note that ζ fixes i, j, b_1, b_2, b_3, b_4, so that ζ is in $N_{\mathbf{w}}$ and $(K)\eta\zeta = L$. Since $\eta\zeta$ is in $N_{\mathbf{w}}$, each orbit of $N_{\mathbf{w}}$ on 0_3 has order $330 \cdot 2^7$ which is, in fact, the order of 0_3. Thus 0_3 is a single orbit for $N_{\mathbf{w}}$ on $\Lambda_2^2(\mathbf{w})$.

The other orbit of $N_{\mathbf{w}}$ on $\Lambda_2^2(\mathbf{w})$ consists of vectors of the form

$$(2\mathbf{v}_K)\epsilon_D$$

where $\{i, j\}$ is contained in K, D is in $R(C)$ and $|D \cap \{i, j\}| = 1$. Denote this collection by 0_4. Note that $0_3 \cup 0_4 =$

$\Lambda_2^2(\mathbf{w})$. We will now show that 0_4 is a single orbit for $N_\mathbf{w}$ on $\Lambda_2^2(\mathbf{w})$.

By Table A2.3, $\{i, j\}$ appears in exactly 77 8-sets of $R(C)$. For each 8-set K, there are 2^7 of the form $(2\mathbf{v}_K)\epsilon_D$, 2^6 of which will be in 0_4. Thus 0_4 has a total of $77 \cdot 2^6 = 4928$ vectors. Note that 0_4 is preserved as a whole by $N_\mathbf{w}$.

For K and L, 8-sets in $R(C)$ containing $\{i, j\}$, the quintuple transitivity of M_{24} implies that there is an element of $N_\mathbf{w}$ which will send $(2\mathbf{v}_K)\epsilon_D$ to $(2\mathbf{v}_L)\epsilon_E$ where D and E are in $R(C)$ and intersect $\{i, j\}$ in exactly one element. Still to be shown is that if D and E are two distinct sets in $R(C)$ that contain exactly one element of $\{i, j\}$, then there is some element of $N_\mathbf{w}$ which moves $(2\mathbf{v}_K)\epsilon_D$ to $(2\mathbf{v}_K)\epsilon_E$.

Suppose first that i is in both D and E. Then both ϵ_D and ϵ_E map $\mathbf{v}_i$ to $-\mathbf{v}_i$ and $\mathbf{v}_j$ to $\mathbf{v}_j$, so that $\epsilon_D\epsilon_E$ is in $N_\mathbf{w}$ and

$$((2\mathbf{v}_K)\epsilon_D)\epsilon_D\epsilon_E = (2\mathbf{v}_K)\epsilon_E.$$

Suppose next that i is in D and j is in E. Pick μ in $N_\mathbf{w}$ that fixes K as a whole but sends i to j and j to i. Then the symmetry $\epsilon_D\mu\epsilon_E$ has the following effect on i and j:

$$i \xrightarrow{\epsilon_D} -i \xrightarrow{\mu} -j \xrightarrow{\epsilon_E} j$$

$$j \xrightarrow{\epsilon_D} j \xrightarrow{\mu} i \xrightarrow{\epsilon_E} i.$$

Thus $\epsilon_D\mu\epsilon_E$ is in $N_\mathbf{w}$ and $((2\mathbf{v}_K)\epsilon_D)\epsilon_D\mu\epsilon_E = (2\mathbf{v}_K)\mu\epsilon_E = (2\mathbf{v}_K)\epsilon_E$. Consequently, 0_4 is a single orbit of $N_\mathbf{w}$ on $\Lambda_2^2(\mathbf{w})$.

Finally, we show that $0_5 = \Lambda_2^3(\mathbf{w})$ is a single orbit. The set 0_5 consists of those points of the form

$$(\mathbf{v}_\Omega - 4\mathbf{v}_k)\epsilon_D,$$

where k is not in $\{i, j\}$ and D is in $R(C)$ such that

$$|D \cap \{i, j\}| = 1.$$

Clearly, these are preserved by $N_\mathbf{w}$.

There are 22 vectors of the form $\mathbf{v}_\Omega - 4\mathbf{v}_k$, k not in $\{i, j\}$. Asking when a set in $R(C)$ intersects $\{i, j\}$ in exactly one element is the same as asking what row vectors in $R(C)$ have differing entries at positions i and j. As noted previously, this happens exactly half the time so that there are 2^{11} such sets. Thus 0_5 has $22 \cdot 2^{11} = 45056$ vectors.

The quintuple, hence single, transitivity of M_{24} shows that there is an element of $N_\mathbf{w}$ which, for given k and l in Ω, sends $(\mathbf{v}_\Omega - 4\mathbf{v}_k)\epsilon_D$ to $(\mathbf{v}_\Omega - 4\mathbf{v}_l)\epsilon_E$. We now show that for given D and E in $R(C)$ that intersect $\{i, j\}$ in a single element, $N_\mathbf{w}$ contains a symmetry that sends $(\mathbf{v}_\Omega - 4\mathbf{v}_k)\epsilon_D$ to $(\mathbf{v}_\Omega - 4\mathbf{v}_k)\epsilon_E$.

If both D and E contain i, then the computations

$$i \xrightarrow{\epsilon_D} -i \xrightarrow{\epsilon_E} i \quad \text{and} \quad j \xrightarrow{\epsilon_D} j \xrightarrow{\epsilon_E} j,$$

show that $\epsilon_D\epsilon_E$ is in $N_\mathbf{w}$, and

$$((\mathbf{v}_\Omega - 4\mathbf{v}_k)\epsilon_D)\epsilon_D\epsilon_E = (\mathbf{v}_\Omega - 4\mathbf{v}_k)\epsilon_E.$$

If i is in D and j in E, let θ be in $N_\mathbf{w}$ so that the permutation θ sends $i \to j$, $j \to i$ and $k \to k$. Let $\pi = \epsilon_D\theta\epsilon_E$. Its effect on i and j is as follows:

$$i \xrightarrow{\epsilon_D} -i \xrightarrow{\theta} -j \xrightarrow{\epsilon_E} j$$

$$j \xrightarrow{\epsilon_E} j \xrightarrow{\theta} i \xrightarrow{\epsilon_E} i.$$

Thus $(i)\pi = j$ and $(j)\pi = i$ so that the symmetry π is in $N_\mathbf{w}$. It is easy to see that $((\mathbf{v}_\Omega - 4\mathbf{v}_k)\epsilon_D)\pi = (\mathbf{v}_\Omega - 4\mathbf{v}_k)\epsilon_E$. Similar cases occur when i and j are reversed. The result is that 0_5 is a single orbit. The sizes of the orbits 0_1 through 0_5 are given, modulo 23, in Table 3.18.

Recall that the orbits of $H_\mathbf{w}$ on $\Lambda_2(\mathbf{w})$ all have order divisible by 23 and are unions of those of $N_\mathbf{w}$, i.e., the 0_i. From Table 3.18, it is easy to see that no sum, some or all of the numbers 2, 4, 12, 6, 22, other than $2 + 4 + 12 + 6 + 22$ is divisible by 23. The set $\Lambda_2(\mathbf{w})$ is a single orbit for $H_\mathbf{w}$. Therefore, $H_\mathbf{x}$ is transitive on $\Lambda_2(\mathbf{x})$ for any $\mathbf{x}$ in Λ_2. ■

TABLE 3.18

Orbit	Size	Size (mod 23)
0_1	2	2
0_2	924	4
0_3	42,240	12
0_4	4,928	6
0_5	45,056	22

The orbits of $N_{\mathbf{w}}$ on $\Lambda_2(\mathbf{w})$.

THEOREM 3.9. $|\cdot 0| = 2^{22}\cdot 3^9\cdot 5^4\cdot 7^2\cdot 11\cdot 13\cdot 23.$

Since any $\mathbf{y}$ in $\Lambda_2(\mathbf{x})$ is orthogonal to $\mathbf{x}$, $|\Lambda_2(\mathbf{x})| = [H_{\mathbf{x}}:H_{\mathbf{x},\mathbf{y}}]$ where $H_{\mathbf{x},\mathbf{y}}$ is the set of symmetries of H fixing both $\mathbf{x}$ and $\mathbf{y}$. Similarly, $|\Lambda_2| = [H:H_{\mathbf{x}}]$ since H is transitive on Λ_2. Table 3.18 shows that $|\Lambda_2(\mathbf{x})| = 93150$; thus

$$\begin{aligned} |H| &= [H:H_{\mathbf{x}}][H_{\mathbf{x}}:H_{\mathbf{x},\mathbf{y}}]|H_{\mathbf{x},\mathbf{y}}| \\ &= 196560\cdot 93150\cdot |H_{\mathbf{x},\mathbf{y}}| \\ &= 2^5\cdot 3^7\cdot 5^3\cdot 7\cdot 13\cdot 23\cdot |H_{\mathbf{x},\mathbf{y}}|. \end{aligned}$$

This means that the order of H is determined by the order of $H_{\mathbf{x},\mathbf{y}}$.

Since the remarks in the last paragraph hold for any orthogonal pair $\mathbf{x}$ and $\mathbf{y}$ in Λ_2, they hold in particular for $\mathbf{x} = 4\mathbf{v}_i + 4\mathbf{y}_j$ and $\mathbf{y} = 4\mathbf{v}_i - 4\mathbf{v}_j$. Any λ fixing both $\mathbf{x}$ and $\mathbf{y}$ must fix $\mathbf{v}_i$. But by Theorem 3.1, such symmetries are, in fact, in N, so that $H_{\mathbf{x},\mathbf{y}}$ is a subgroup of N. Furthermore, if $\mathbf{v}_i$ is fixed by λ, so is $\mathbf{v}_j$, so that $H_{\mathbf{x},\mathbf{y}}$ is the subgroup of N fixing the coordinate vectors $\mathbf{v}_i$ and $\mathbf{v}_j$. We will now show that $H_{\mathbf{x},\mathbf{y}}$ is a group of the form $E_{10}M_{22}$.

Now, any element of N is of the form $\epsilon_D\pi$ where ϵ_D is in E_{12} and π is in M_{24}. Furthermore, if $\epsilon_E \neq \epsilon_D$, then $\epsilon_E\pi \neq \epsilon_D\pi_2$ for any π_1, π_2 in M_{24}. Thus

$$|N| = |E_{12}||M_{24}| = 2^{12}\cdot |M_{24}|.$$

If λ in N fixes both $\mathbf{v}_i$ and $\mathbf{v}_j$, and $\lambda = \epsilon_D \pi$, both ϵ_D and π must fix $\mathbf{v}_i$ and $\mathbf{v}_j$. This means that π must be in a copy of M_{22} and D contains neither i nor j.

How many such D are there? Since the elements of Ω are distributed evenly over the sets of $R(C)$, half of the sets of $R(C)$ miss i and half the remaining also miss j. Thus $\{i, j\}$ misses 2^{10} sets of $R(C)$ and $|H_{\mathbf{x},\mathbf{y}}| = 2^{10} \cdot |M_{22}|$. Appendix 4 shows that M_{22} has order $22 \cdot 21 \cdot 20 \cdot 48$; thus $H_{\mathbf{x},\mathbf{y}}$ has order $2^{10} \cdot 2^7 \cdot 3^2 \cdot 5 \cdot 7 \cdot 11$. We, therefore, have that H has order

$$2^{22} \cdot 3^9 \cdot 5^4 \cdot 7^2 \cdot 11 \cdot 13 \cdot 23 = 8{,}315{,}553{,}613{,}086{,}720{,}000.$$

Since the order of H is completely determined by the assumption that H is a subgroup of $\cdot 0$ strictly containing N, H must be $\cdot 0$. ■

COROLLARY 3.10. *N is a maximal subgroup of* $\cdot 0$.

COROLLARY 3.11. $\cdot 0$ *is generated by* α, β, γ, δ, ϵ *and* ζ.

Let us review briefly what we have found. The group, $\cdot 0$, of Euclidean congruences of Leech's lattice, Λ, which also fix the origin (equivalently, the symmetries of the 24-dimensional polytope whose 196560 vertices are those points of Λ at distance $4\sqrt{2}$ from the origin), is an extremely large group containing isomorphic copies of the sporadic simple groups M_{24}, M_{23} and M_{22} and has a particularly nice maximal subgroup $N = E_{12}M_{24}$. However, this is not all. As it turns out, $\cdot 0$ contains several other simple groups besides those of Mathieu, two of them new, $\cdot 2$ and $\cdot 3$. Moreover, the quotient of $\cdot 0$ by its center is a third new simple group, $\cdot 1$. These new simple groups, $\cdot 1$, $\cdot 2$ and $\cdot 3$ are also called Co_1, Co_2 and Co_3, respectively [**65**, p. 709]. Like the Mathieu groups, they belong to no known infinite family of simple groups and are thus called "sporadic."

We will show the simplicity of $\cdot 1$. Before doing this, however, we wish to make one more remark concerning Conway's work in [24]. In that paper, Conway assumed that H is a subgroup of $\cdot 0$, strictly containing N, and derived its order, thus showing that $H = \cdot 0$. He used the fact that H is transitive on the vertices of P so that $|H| = 196560 \cdot |H_{\mathbf{x}}|$, where $H_{\mathbf{x}}$ is the subgroup of H fixing the vertex $\mathbf{x}$. Furthermore, he showed that $H_{\mathbf{x}}$ is transitive on $\Lambda_2(\mathbf{x})$, the subset of Λ_2 consisting of those vertices perpendicular to $\mathbf{x}$. This meant that

$$|H| = 196560 \cdot [H_{\mathbf{x}} : H_{\mathbf{x},\mathbf{y}}]\,|H_{\mathbf{x},\mathbf{y}}| = 196560 \cdot 93150 \cdot |H_{\mathbf{x},\mathbf{y}}|$$

where $\mathbf{y}$ is perpendicular to $\mathbf{x}$ and $H_{\mathbf{x},\mathbf{y}}$ is the subgroup of H fixing both $\mathbf{x}$ and $\mathbf{y}$. In his paper, he then showed that $H_{\mathbf{x},\mathbf{y}}$ is of the form $E_{10}M_{22}$ and so he was able to deduce the order of H. This step, however, differs from the argument in his original discovery. Said Conway, "We made two distinct refinements.... The first which I definitely remember is the matrix." [27] This has already been dealt with in this section.

Conway had a little more difficulty recalling the second refinement:

> I showed that if there was any symmetry at all, it had to be transitive. Then, again, if there is any new symmetry, the group has to be transitive on pairs of touching spheres and so on.... I eventually showed that I had a particular set of five spheres, or vectors, whatever you like to call them, and that if there was any new symmetry [outside N], then there has to be a symmetry taking these five vectors to these five in this particular order. And by this time, it's beginning to get pretty restricted, you see, because the number of spheres that touch all five is getting quite small.... Eventually you get down to a case where there is a unique sphere doing such and such.... In this way I was in a position to say that I could work out exactly what the group order was to within a factor of two. Basically, I got down to a certain stage and I could tell that the thing had at most two images under what was left of the group. [26]
>
> So you have exactly what the group was. Knowing the number of possibilities at each level, you get the big product of that number by the various numbers at each stage. Now, in at least one of the papers [24] that became refined to an argument that said if you fix two orthogonal vectors, then you're already in this subgroup

> $[E_{10}N_{22}]$ of known order, which is, in principle, sort of the same thing, but it looks a lot more elegant. Of course, it's typical that the first time you do things is never a very elegant time. [**27**]

Section 5. New Simple Groups

One thing Conway did not do at the time he and Thompson worked on the order and structure of $\cdot 0$ was to show the simplicity of the three new simple groups—$\cdot 1$, $\cdot 2$ and $\cdot 3$.

> I never did that, to tell you the truth, because I always regarded it as sort of too obvious to need proof. Not literally, but I wasn't really concerned with a formal proof. Thompson had already done it, sort of instantly. Later, at a lecture at Oxford, I was really quite flabbergasted. Somebody asked me, 'How do you know this group [$\cdot 1$] is simple?' And I just said that for some reason it had never occurred to me to prove that it was simple. I said that it should be obvious. Then somebody, Peter Neumann, I think, very kindly suggested a proof which is given in one of the papers. [**26**] He just sketched the proof there and then. I mean, the thing really was trivial; so, I think, my intuition was right, that it wasn't an important thing. But I wasn't capable of accomplishing that triviality, just because I wasn't knowledgeable in group theory at that time. I didn't actually follow his argument because I didn't know some of the technical terms involved. [**27**]

We now give Neumann's proof that $\cdot 1$ is simple. Recall that $\cdot 1$ is defined to be the quotient of $\cdot 0$ by its center.

Our first task is to show that $\cdot 0$ has only a two element center which consists of just the identity of $\cdot 0$, 1, and its negative, -1, which sends $\mathbf{v}_i$ to $-\mathbf{v}_i$ for each i in Ω.

LEMMA 3.12. *The center of* $\cdot 0$ *has order two.*

Let $\lambda \neq \pm 1$ be in $\cdot 0$. We will show that λ is not in the center. Recall that $\cdot 0$ contains the symmetry α corresponding to the permutation $(\infty)(0\ 1\ 2\ \ldots\ 22)$ and also contains γ, which has the 2-cycle $(\infty 0)$. Clearly α, as a permutation, fixes ∞ and moves all the other elements of Ω. A permutation which fixes $i \neq \infty$ and moves all the other elements of Ω is

$$\theta = \alpha^{23-i}\gamma\alpha\gamma\alpha^{i},$$

and it has the same cycle structure as α. If λ sends $\mathbf{v}_i$ to $\mathbf{v}_i$ or $-\mathbf{v}_i$ for all i in Ω, pick j such that λ sends $\mathbf{v}_j$ to $-\mathbf{v}_j$ and $\mathbf{v}_{j+1}$ to $\mathbf{v}_{j+1}$ $(22+1=0)$. Such a j must exist since $\lambda \neq \pm 1$. Thus

$$(\mathbf{v}_j)\lambda\alpha = (-\mathbf{v}_j)\alpha = -\mathbf{v}_{j+1}$$

while

$$(\mathbf{v}_j)\alpha\lambda = (\mathbf{v}_{j+1})\lambda = \mathbf{v}_{j+1},$$

so that λ does not commute with α.

If λ sends $\mathbf{v}_i$ to $\pm\mathbf{v}_j$, for some $j \neq i$, then

$$(\mathbf{v}_i)\lambda\theta = (\pm\mathbf{v}_j)\theta = \pm\mathbf{v}_k \neq \pm\mathbf{v}_j$$

for appropriate θ, while

$$(\mathbf{v}_i)\theta\lambda = (\mathbf{v}_i)\lambda = \pm\mathbf{v}_j.$$

Again, λ is not in the center of $\cdot 0$. Finally, suppose that $(\mathbf{v}_i)\lambda = \mathbf{w} \neq \pm\mathbf{v}_j$ for any j in Ω. In particular $(8\mathbf{v}_i)\lambda$ must be in the subset Λ_4 of Λ. Since the sum of the squares of the coordinates of any vector in Λ_4 is 64, only the following shapes could occur:

$$(0^{23}, \pm 8), (0^{20}, \pm 4^4), (0^{16}, \pm 2^7, \pm 6), (0^{14}, \pm 2^8, \pm 4^2),$$

$$(0^{11}, \pm 2^{12}, \pm 4), (0^8, \pm 2^{16}), (\pm 1^{21}, \pm 3^2, \pm 5), (\pm 1^{19}, \pm 3^5).$$

The only vector in the above family which θ can fix is $8\mathbf{v}_i$. Thus

$$(\mathbf{v}_i)\lambda\theta = (\mathbf{w})\theta = \mathbf{x} \neq \mathbf{w}$$

while

$$(\mathbf{v}_i)\theta\lambda = (\mathbf{v}_i)\lambda = \mathbf{w}.$$

This shows that λ is not in the center of $\cdot 0$. In other words, the center of $\cdot 0$ is $\{1, -1\}$ and $\cdot 1$ is a group of order

$$2^{21} \cdot 3^9 \cdot 5^4 \cdot 7^2 \cdot 11 \cdot 13 \cdot 23. \qquad \blacksquare$$

To show that $\cdot 1$ is simple, it suffices to prove the following theorem.

THEOREM 3.13. *The group $\cdot 0$ is transitive on ordered pairs of vectors of Λ_2 with any given scalar product.*

(Recall that Λ_2 is the set of $\mathbf{v}$ in Λ such that $\mathbf{v} \cdot \mathbf{v} = 16 \cdot 2$, that is, the set of 196560 vertices of P).

What scalar products do we have to consider? Since $\cdot 0$ preserves scalar products and since $\cdot 0$ is transitive on Λ_2 by Theorem 3.7, it will be necessary to check only the inner product say, of $4\mathbf{v}_i + 4\mathbf{v}_j$ with the other elements of Λ_2. All possible types are listed in Table 3.19.

TABLE 3.19

Vector Type	Product
$\pm(4\mathbf{v}_i + 4\mathbf{v}_j)$	± 32
$\pm(4\mathbf{v}_i - 4\mathbf{v}_j)$	0
$\pm 4\mathbf{v}_l \pm 4\mathbf{v}_k$, $k \neq j$, $l \in \{i, j\}$	± 16
$\pm 4\mathbf{v}_h \pm 4\mathbf{v}_k$, h, i, j, k, all distinct	0
$\pm(2\mathbf{v}_i + 2\mathbf{v}_j + \cdots)$	± 16
$\pm(2\mathbf{v}_i - 2\mathbf{v}_j + \cdots)$	0
$\pm 2\mathbf{v}_i \pm 2\mathbf{v}_k + \cdots$	± 8
$\pm 2\mathbf{v}_h \pm 2\mathbf{v}_k + \cdots$	0
$\pm(3\mathbf{v}_i + \mathbf{v}_j + \cdots)$	± 16
$\pm(3\mathbf{v}_i - \mathbf{v}_j + \cdots)$	± 8
$\pm(\mathbf{v}_i + \mathbf{v}_j + \cdots)$	± 8
$\pm(\mathbf{v}_i - \mathbf{v}_j + \cdots)$	0

The scalar products of $4\mathbf{v}_i + 4\mathbf{v}_j$ with vector in Λ_2.

This shows that the only scalar products are 0, ± 8, ± 16 and ± 32.

Now define $\Lambda_2(\mathbf{x}, m)$ to be the subset of Λ_2 consisting of those $\mathbf{y}$ such that $(\mathbf{x}, \mathbf{y}) = m$. For instance, $\Lambda_2(\mathbf{x}, 0) = \Lambda_2(\mathbf{x})$ and $\Lambda_2(\mathbf{x}, m) = \emptyset$ if $m \neq 0, \pm 8, \pm 16$ or ± 32. Suppose that it can be shown that $\cdot 0_{\mathbf{x}}$ is transitive on $\Lambda_2(\mathbf{x}, m)$. If $(\mathbf{x}, \mathbf{y}) = m = (\mathbf{v}, \mathbf{w})$, then there is an η in $\cdot 0$ (by Theorem 3.7) which sends $\mathbf{x}$ to $\mathbf{v}$ and a ξ in $\cdot 0_{\mathbf{v}}$ which sends $(\mathbf{y})\eta$ to $\mathbf{w}$, so that

$$(\mathbf{x})\eta\xi = (\mathbf{v})\xi = \mathbf{v}$$

and

$$(\mathbf{y})\eta\xi = ((\mathbf{y})\eta)\xi = \mathbf{w}.$$

Then $\cdot 0$ would be transitive on ordered pairs of vectors in Λ_2 with the same scalar product. Of course, we must show that $\cdot 0_{\mathbf{x}}$ is transitive on $\Lambda_2(\mathbf{x}, m)$.

Suppose $m = 32$. Table 3.19 shows that $\Lambda_2(\mathbf{x}, 32) = \{\mathbf{x}\}$ and thus $|\Lambda_2(\mathbf{x}, 32)| = 1$. If $m = -32$, $\Lambda_2(\mathbf{x}, -32) = \{-\mathbf{x}\}$ and again $|\Lambda_2(\mathbf{x}, -32)| = 1$. These are the two trivial cases.

There are, from Table 3.19, three types of vectors in $\Lambda_2(\mathbf{x}, 16)$; they are listed separately in Table 3.20.

TABLE 3.20

Vector Type	Number
$4\mathbf{v}_i \pm 4\mathbf{v}_k$ or $4\mathbf{v}_j \pm 4\mathbf{v}_k$, $k \neq i, j$	$2 \cdot 2 \cdot 22 = 88$
$(2\mathbf{v}_K)\epsilon_D$, $\{i, j\} \subseteq K$, $\{i, j\} \cap D = \emptyset$	$77 \cdot 2^5 = 2{,}464$
$(\mathbf{v}_\Omega - 4\mathbf{v}_i)\epsilon_D$, i in D, j not in D or reverse i and j	$2 \cdot 2^{10} = 2{,}048$

The vectors in $\Lambda_2(\mathbf{x}, 16)$.

We first find the orbits of $N_{\mathbf{x}}$ on $\Lambda_2(\mathbf{x}, 16)$ and then show that $\Lambda_2(\mathbf{x}, 16)$ is a single orbit of $\cdot 0_{\mathbf{x}}$.

The vector types in Table 3.20 will be designated by the symbols 0_1^{16}, 0_2^{16} and 0_3^{16}, respectively. Each of the 0_i^{16} is fixed, as a whole, by $N_{\mathbf{x}}$, so that the orbits of $N_{\mathbf{x}}$ on $\Lambda_2(\mathbf{x}, 16)$ are subsets of the 0_i^{16}.

Since M_{24} is in N, there is an element in $N_{\mathbf{x}}$ that sends $\mathbf{v}_i$ to

$\mathbf{v}_j$ and vice versa. The group $N_{\mathbf{x}}$ will be transitive on 0_1^{16} if it has an element that sends $4\mathbf{v}_i \pm 4\mathbf{v}_k$ to $4\mathbf{v}_i \pm 4\mathbf{v}_l$, k and l not in $\{i, j\}$. Suppose $4\mathbf{v}_i - 4\mathbf{v}_k$ is to be sent to $4\mathbf{v}_i + 4\mathbf{v}_l$. (The other cases are done similarly.) Pick an 8-set in $R(C)$ missing $\{i, j\}$ but containing k. Further, pick η in M_{24} sending i to i, j to j and k to l. Then both ϵ_K and η are in $N_{\mathbf{x}}$ and

$$(4\mathbf{v}_i - 4\mathbf{v}_k)\epsilon_K\eta = (4\mathbf{v}_i + 4\mathbf{v}_k)\eta = 4\mathbf{v}_i + 4\mathbf{v}_l.$$

Thus 0_1^{16} is a single orbit for $N_{\mathbf{x}}$ on $\Lambda_2(\mathbf{x}, 16)$.

The quintuple transitivity of M_{24} implies that there is, in $N_{\mathbf{x}}$, an element ξ that sends K to L, where both K and L are 8-sets in $R(C)$, since $\{i, j\}$ is contained in both K and L. Also, ϵ_D and ϵ_E are both in $N_{\mathbf{x}}$ if $D \cap \{i, j\} = E \cap \{i, j\} = \emptyset$. Thus

$$((2\mathbf{v}_K)\epsilon_D)\epsilon_D\xi\epsilon_E = (2\mathbf{v}_K)\xi\epsilon_E = (2\mathbf{v}_L)\epsilon_E,$$

and so 0_2^{16} is also a single orbit for $N_{\mathbf{x}}$ on $\Lambda_2(\mathbf{x}, 16)$.

If both D and E intersect $\{i, j\}$ only in i, $\epsilon_D\epsilon_E$ is in $N_{\mathbf{x}}$ and

$$((\mathbf{v}_\Omega - 4\mathbf{v}_i)\epsilon_D)\epsilon_D\epsilon_E = (\mathbf{v}_\Omega - 4\mathbf{v}_i)\epsilon_E.$$

It is almost as easy to find an element of $N_{\mathbf{x}}$ which sends $(\mathbf{v}_\Omega - 4\mathbf{v}_i)\epsilon_D$ to $(\mathbf{v}_\Omega - 4\mathbf{v}_j)\epsilon_F$ where j is in F, but i is not. Let λ be an element in $N_{\mathbf{x}}$ that interchanges $\mathbf{v}_i$ and $\mathbf{v}_j$. Then $\epsilon_D\lambda\epsilon_F$ interchanges $\mathbf{v}_i$ and $\mathbf{v}_j$ and so is in $N_{\mathbf{x}}$, whence

$$((\mathbf{v}_\Omega - 4\mathbf{v}_i)\epsilon_D)\epsilon_D\lambda\epsilon_F = (\mathbf{v}_\Omega - 4\mathbf{v}_j)\epsilon_F.$$

Of course, this means that 0_3^{16} is also a single orbit.

The mapping α moves every element of Λ_2 except $\pm(\mathbf{v}_\Omega - 4\mathbf{v}_\infty)$ and is in both $N_{\mathbf{x}}$ and $\cdot 0_{\mathbf{x}}$ if $\mathbf{x} = \mathbf{v}_\Omega - 4\mathbf{v}_\infty$. For this $\mathbf{x}$, the orbits of $\cdot 0_{\mathbf{x}}$ on $\Lambda_2(\mathbf{x}, 16)$ are, as before, each divisible by 23 since α has order 23. Because $\cdot 0$ is transitive on Λ_2, the orders of the orbits of $\cdot 0_{\mathbf{x}}$ on $\Lambda_2(\mathbf{x}, 16)$ are also divisible by 23 when $\mathbf{x} = 4\mathbf{v}_i + 4\mathbf{v}_j$.

Since $|0_1^{16}| \equiv 19 \pmod{23}$, $|0_2^{16}| \equiv 3$ and $|0_3^{16}| \equiv 1$, $\Lambda_2(\mathbf{x}, 16)$ is a single orbit for $\cdot 0_{\mathbf{x}}$ containing

$$|0_1^{16}| + |0_2^{16}| + |0_3^{16}| = 4600 \text{ elements.}$$

A similar argument shows that $\cdot 0_{\mathbf{x}}$ has a single orbit on $\Lambda_2(\mathbf{x}, -16)$, also containing 4600 elements.

We now consider $\Lambda_2(\mathbf{x}, 8)$. From Table 3.19 there are again three vector types, shown in Table 3.21.

TABLE 3.21

Vector Type	Number
$(2\mathbf{v}_K)\epsilon_D$, i in K but not in D, j not in K (or the roles of i and j reversed)	$2\cdot(253-77)\cdot 2^6 = 22{,}528$
$(4\mathbf{v}_i - \mathbf{v}_\Omega)\epsilon_D$, $\{i, j\} \cap D = \emptyset$	$2\cdot 2^{10} = 2{,}048$
$\{\mathbf{v}_\Omega - 4\mathbf{v}_k)\epsilon_D$, $k \neq i, j$, $\{i, j\} \cap D = \emptyset$	$22\cdot 2^{10} = 22{,}528$

The vectors in $\Lambda_2(\mathbf{x}, 8)$, $\mathbf{x} = 4\mathbf{v}_i + 4\mathbf{v}_j$.

Label the three sets 0_1^8, 0_2^8 and 0_3^8, respectively.

The sets 0_2^8 and 0_3^8 are easily shown to be single orbits for $N_{\mathbf{x}}$. Suppose, for 0_1^8, i is in neither D nor E. If j is in either $D \cap E$ or $\Omega - (D \cup E)$, $\epsilon_D\epsilon_E$ is in $N_{\mathbf{x}}$ and $((2\mathbf{v}_K)\epsilon_D)\epsilon_D\epsilon_E = (2\mathbf{v}_K)\epsilon_E$. If j is in D but not E, pick L, $j \in L$, an 8-set disjoint from K. Then $\epsilon_D\epsilon_L\epsilon_E$ is in $N_{\mathbf{x}}$ and $((2\mathbf{v}_K)\epsilon_D)\epsilon_D\epsilon_L\epsilon_E = (2\mathbf{v}_K)\epsilon_E$. Since M_{24} has permutations with structure 2^{12} (e.g., γ), $N_{\mathbf{x}}$ contains a symmetry of $\cdot 0$ that sends $2\mathbf{v}_K$ to $2\mathbf{v}_M$, where j is in M but i is not. Suppose i is in both K and L, where K and L are 8-sets in $R(C)$, but j is not. If $N_{\mathbf{x}}$ contains a symmetry that sends $2\mathbf{v}_K$ to $2\mathbf{v}_L$, 0_1^8 is a single orbit for $N_{\mathbf{x}}$ on $\Lambda_2(\mathbf{x}, 8)$. We now show that such a symmetry indeed exists.

Let M be an 8-set in $R(C)$ that contains both i and j and some other three elements of K, say a_1, a_2, a_3. Let η in M_{24} send i to i, j to j, a_1 to b_1, a_2 to b_2 and a_3 to b_3, where b_1, b_2 and b_3 are distinct from i, in L. If $(K)\eta = L$, we are done, since, by definition, η is in $N_{\mathbf{x}}$. Otherwise, $(K)\eta \cap L = \{i, b_1, b_2, b_3\}$. Let $(K)\eta$ contain the additional element a_4, and L contain the additional element a_5. By Lemma A4.3, the group of symmetries fixing $(M)\eta$ elementwise contains a symmetry ζ that sends a_4 to a_5. Since ζ fixes i, j, b_1, b_2 and b_3, ζ, as a symmetry, is in $N_{\mathbf{x}}$ and $(\mathbf{v}_K)\eta\zeta = \mathbf{v}_L$.

As in the previous case, 23 divides the orders of the orbits of $\cdot 0_{\mathbf{x}}$ on $\Lambda_2(\mathbf{x}, 8)$. Since $22528 \equiv 11 \pmod{23}$ and $2048 \equiv 1$, $\Lambda_2(\mathbf{x}, 8)$ is a single orbit for $\cdot 0_{\mathbf{x}}$ containing 47104 elements. Likewise, $\Lambda_2(\mathbf{x}, -8)$ is a single orbit for $\cdot 0_{\mathbf{x}}$ containing 47104 elements.

The last case, $\Lambda_2(\mathbf{x}, 0) = \Lambda_2(\mathbf{x})$, is covered by Theorem 3.8. Table 3.18 shows that $\Lambda_2(\mathbf{x})$ has 93150 elements. Since $\cdot 0_{\mathbf{x}}$ is transitive on each of the sets $\Lambda_2(\mathbf{x}, m)$, the theorem is proved. ■

Call a pair $\{\mathbf{x}, -\mathbf{x}\}$, $\mathbf{x}$ in Λ_2, a *diameter* and denote by $\overline{\Lambda_2}$ the set of 98280 diameters of Λ_2. This is easy to picture geometrically since any such pair lies on a line, passing through the origin in E^{24}, which is a diameter for the polytope P of 196560 vertices.

For λ in $\cdot 0$, the corresponding element of $\cdot 1$ is $\{\lambda, -\lambda\}$, which acts naturally on $\overline{\Lambda_2}$:

$$(\{\mathbf{x}, -\mathbf{x}\})\{\lambda, -\lambda\} = \{(\mathbf{x})\lambda, -(\mathbf{x})\lambda\}.$$

Let $\cdot 1_{\{\mathbf{x}, -\mathbf{x}\}}$ be the stabilizer in $\cdot 1$ of $\{\mathbf{x}, -\mathbf{x}\}$, that is, the set of all elements in $\cdot 1$ which fix the diameter $\{\mathbf{x}, -\mathbf{x}\}$. Since $\cdot 0$ is transitive on ordered pairs of elements of Λ_2 with any given scalar product by Theorem 3.13, there is a μ in $\cdot 0_{\mathbf{x}}$ which sends $\mathbf{y}$ to $\mathbf{z}$ where $(\mathbf{x}, \mathbf{y}) = (\mathbf{x}, \mathbf{z}) = m$. The corresponding $\{\mu, -\mu\}$ in $\cdot 1_{\{\mathbf{x}, -\mathbf{x}\}}$ sends $\{\mathbf{y}, -\mathbf{y}\}$ to $\{\mathbf{z}, -\mathbf{z}\}$. In particular, $\cdot 1_{\{\mathbf{x}, -\mathbf{x}\}}$ has orbits of orders 1, 4600, 47104 and 46575 on $\overline{\Lambda_2}$.

LEMMA 3.14. *The group* $\cdot 1$ *acts primitively on* $\overline{\Lambda_2}$.

A permutation group G is *imprimitive* if the set of letters can be divided into a finite family of disjoint subsets S_1, S_2, $\ldots$, S_m so that every permutation of G either permutes the elements of an S_i among themselves or sends S_i to another S_j.

The trivial cases when the family consists of only one set S, or of singletons, are excepted. The sets S_i are called the *sets of imprimitivity*. An example is

$$G = \{1, (12)(34), (13)(24), (14)(23)\}.$$

One family of sets of imprimitivity consists of $S_1 = \{1, 2\}$ and $S_2 = \{3, 4\}$. If G is not imprimitive, G is called *primitive*. Note that if G is transitive, $|S_i| = |S_j|$ for all i and j.

Each element of $\cdot 1$ permutes the 98280 diameters. Since $\cdot 0$ is transitive on Λ_2, $\cdot 1$ is transitive on the set of diameters. Denote the elements of $\overline{\Lambda_2}$ by $\mathbf{x}$ instead of $\{\mathbf{x}, -\mathbf{x}\}$. Suppose $\cdot 1$ is imprimitive. The transitivity of $\cdot 1$ implies that $|S_i|$ is a divisor of 98280. Let $\mathbf{x}$ be in S_1. Since $|S_1| > 1$, there is a $\mathbf{y}$ in S_1 whose orbit under $\cdot 1_{\mathbf{x}}$ has order 4600, 47104 or 46575. Since $\cdot 1_{\mathbf{x}}$ fixes $\mathbf{x}$, S_1 must be sent to itself by $\cdot 1_{\mathbf{x}}$ and so $|S_1|$ is at least as large as 4601, 47105 or 46576. Since none of these numbers divides 98280, S_1 must contain an element $\mathbf{z}$ outside the orbit of $\mathbf{y}$ and different from $\mathbf{x}$. But then S_1 contains at least $1 + 4600 + 46575 = 51176$ elements and so must be all of $\overline{\Lambda_2}$. This contradicts the fact that $\cdot 1$ was assumed to be imprimitive. Thus $\cdot 1$ acts primitively on $\overline{\Lambda_2}$ and Lemma 3.14 is proved. ■

We need three results from group theory before proceeding with the proof of the simplicity of $\cdot 1$.

Recall that the *normalizer* of a subgroup H of G, written $N_G(H)$, is the subgroup $\{g \mid g^{-1}Hg = H\}$ of G.

LEMMA 3.15. *Let H be a subgroup of G. Then the number of distinct conjugates of H in G is* $[G : N_G(H)]$.

Argument: We have $g_1^{-1}Hg_1 = g_2^{-1}Hg_2 \Leftrightarrow g_2g_1^{-1}H = Hg_2g_1^{-1} \Leftrightarrow g_2g_1^{-1}$ is in $N_G(H) \Leftrightarrow g_2$ is in $N_G(H)g_1$ and g_1 is in $N_G(H)g_2 \Leftrightarrow N_G(H)g_1 = N_G(H)g_2$ because two right cosets are either disjoint or equal. Thus the map $g^{-1}Hg \mapsto N_G(H)g$ is a bijection and the lemma is proved. ■

If A and B are subgroups of a group G, denote by $A \vee B$ the smallest subgroup containing both A and B. This is the *subgroup generated by A and B.*

LEMMA 3.16. *If A and B are subgroups of a finite group G, then $[A \vee B : B] \geq [A : A \cap B]$.*

Argument: Let $A \cap B = D$. Then A is the disjoint union of the right cosets $Dg_1, Dg_2, \ldots, Dg_r$, where g_i is in A. Consider the corresponding cosets $Bg_1, Bg_2, \ldots, Bg_r$ of B in $A \vee B$. The lemma will be established if we can show that these are distinct.

Suppose $Bg_i = Bg_j$. Then $g_i = bg_j$ for some b in B. Since g_i is in A, so is bg_j and thus b is since g_j is also in A. But then b is in D. This means that bg_j is in Dg_j. Thus $Dg_i = Dg_j$, a contradiction. ■

LEMMA 3.17. *If $|G| = pq$, $p < q$, p and q primes and if p does not divide $q - 1$, then G is cyclic.*

Argument: By Sylow's theorems it is easy to show that G has normal subgroups of order p and q. It is also not difficult to show that the elements in one such subgroup commute with those in the other. But then G has an element of order pq. ■

We are now ready to show the simplicity of $\cdot 1$.

THEOREM 3.18. *The group $\cdot 1$ is simple.*

If $\cdot 1$ is not simple, $\cdot 0$ contains a proper normal subgroup H properly containing its center. If H is not transitive on $\overline{\Lambda}_2$, there exist $\mathbf{x}$ $(=\{\mathbf{x}, -\mathbf{x}\})$ and $\mathbf{y}$ in $\overline{\Lambda}_2$ such that $(\mathbf{x})\eta \neq \mathbf{y}$ for any η in H. Thus H must have at least two orbits in $\overline{\Lambda}_2$. Since $\cdot 0$ is transitive on $\overline{\Lambda}_2$, and since H is normal in $\cdot 0$, the orbits of H in $\overline{\Lambda}_2$ are equinumerous. Furthermore, they are nontrivial since H strictly contains $\{1, -1\}$. Let ζ $(=\{\zeta, -\zeta\})$ in $\cdot 1$ map $\mathbf{x}$ to $\mathbf{z}$ and suppose λ in H sends $\mathbf{x}$ to $\mathbf{w}$ in the same orbit of $\overline{\Lambda}_2$. Then $(\mathbf{w})\zeta = ((\mathbf{x})\lambda)\zeta = (\mathbf{z})\zeta^{-1}\lambda\zeta$ must be in the orbit of $\mathbf{z}$ since $\zeta^{-1}\lambda\zeta$ is in H. This is true whether or not $\mathbf{x}$ and $\mathbf{z}$ are in the same orbits. This means that $\cdot 1$ preserves the orbits of H in $\overline{\Lambda_2}$, so that they are sets of imprimitivity for $\cdot 1$ on $\overline{\Lambda_2}$. This is impossible by Lemma 3.14. Thus H must be transitive on $\overline{\Lambda_2}$. Then, for any $\mathbf{x}$ in $\overline{\Lambda_2}$, $[H:H_{\mathbf{x}}] = |\overline{\Lambda_2}| = 98280 = 13 \cdot 7560$ shows that 13 divides the order of H.

Let P be a 13-Sylow subgroup of H. Then $|P| = 13^k$, and 13^{k+1} does not divide the order of H. Since H is normal in $\cdot 0$, every conjugate, R, of P ($R = \lambda^{-1}P\lambda$ for some λ in $\cdot 0$) in $\cdot 0$ lies in H and so, by Sylow's second theorem [**62**, p. 81], is conjugate to P in H. Thus P has as many conjugates in H as it does in $\cdot 0$.

Let $N_{\cdot 0}(P) = \{\lambda \text{ in } \cdot 0 \mid \lambda^{-1}P\lambda = P\}$ be the normalizer of P in $\cdot 0$. By Lemma 3.15, the number of distinct conjugates of P in $\cdot 0$ is $[\cdot 0:N_{\cdot 0}(P)]$. Similarly, the number of distinct conjugates of P in H is $[H:N_H(P)]$. Since all the distinct conjugates of P in $\cdot 0$ lie also in H, we must have $[\cdot 0:N_{\cdot 0}(P)] = [H:N_H(P)]$. Since

$$[\cdot 0:N_H(P)] = [\cdot 0:H][H:N_H(P)]$$
$$= [\cdot 0:N_{\cdot 0}(P)][N_{\cdot 0}(P):N_H(P)],$$
$$[\cdot 0:H] = [N_{\cdot 0}(P):N_H(P)].$$

But $N_H(P) = N_{\cdot 0}(P) \cap H$ so that

$$[\cdot 0:H] = [N_{\cdot 0}(P):N_{\cdot 0}(P) \cap H]. \qquad (3.22)$$

We now use Lemma 3.16 to show that

$$[N_{\cdot 0}(P) \vee H:H] \geq [N_{\cdot 0}(P):N_{\cdot 0}(P) \cap H].$$

By Equation 3.22,

$$[N_{\cdot 0}(P) \vee H:H] \geq [\cdot 0:H].$$

Since $N_{\cdot 0}(P)$ is in $\cdot 0$, $\cdot 0 = N_{\cdot 0}(P)H$. Now $N_{\cdot 0}(P) \vee H$ is the smallest subgroup of $\cdot 0$ containing $N_{\cdot 0}(P)H$, and thus is $N_{\cdot 0}(P)H$, since H is normal in $\cdot 0$. Thus by [**62**, p. 39],

$$|\cdot 0| = |N_{\cdot 0}(P)H| = \frac{|N_{\cdot 0}(P)||H|}{|N_{\cdot 0}(P) \cap H|}.$$

Since 23 divides the order of $\cdot 0$, it divides either the order of $N_{\cdot 0}(P)$ or the order of H. Suppose the first is true. Let λ be an element of $N_{\cdot 0}(P)$ of order 23. Since 13 divides the order of $\cdot 0$ but no higher power of 13 does, the order of P is 13 and P is generated by a single element μ. Let the group generated by λ be denoted by K. Since $\theta^{-1}P\theta = P$ for each θ in $N_{\cdot 0}(P)$, P is normal in $N_{\cdot 0}(P)$. Thus PK is a subgroup of $N_{\cdot 0}(P)$. But then

$$|PK| = \frac{|P||K|}{|P \cap K|} = |P||K| = 13 \cdot 23.$$

Thus $\cdot 0$ has a subgroup of order $13 \cdot 23$. Such a group is cyclic, by Lemma 3.17, since 13 does not divide 22. This means that $\cdot 0$ has an element of order $13 \cdot 23$, which is impossible by Lemma 3.3. Thus 23 must divide the order of H.

Since H is normal in $\cdot 0$, $H \cap N$ is normal in N. Furthermore, HN is a subgroup of $\cdot 0$. Let $|HN| = 23 \cdot n$. Then

$$23 \cdot n = \frac{|H||N|}{|H \cap N|} = \frac{23 \cdot 23 \cdot m}{|H \cap N|},$$

where 23 divides neither m nor n and so must divide $|H \cap N|$.

We now show that $H \cap N = N$. Since N is maximal by Corollary 3.10, this will imply that $H = N$ and that N must be normal in $\cdot 0$. The proof of Theorem 3.18 will be completed by showing that N is not normal in $\cdot 0$ and thus H, a supposed proper normal subgroup of $\cdot 0$, cannot exist.

To show $H \cap N = N$, we first note that $M_{24}(H \cap N)$ is a subgroup of N, since $H \cap N$ is normal in N. Furthermore,

$$M_{24}(H \cap N) = \frac{|M_{24}||H \cap N|}{|M_{24} \cap H \cap N|}.$$

Since 23 divides the left-hand side and both factors in the numerator on the right-hand side, while no higher power of 23 divides the left-hand side, 23 must divide $|M_{24} \cap H \cap N|$. Thus $M_{24} \cap H \cap N$ is a nontrivial, normal subgroup of M_{24}. Since M_{24} is simple, $H \cap N$ must contain M_{24}.

Let K and L be disjoint 8-sets in $R(C)$ and let η in M_{24} send K to L. The normality of $H \cap N$ in N implies that $\epsilon_K \eta \epsilon_K$ is in $H \cap N$. But then so is $\eta^{-1}\epsilon_K\eta\epsilon_K$. Since $\eta^{-1}\epsilon_K\eta = \epsilon_L$, $\epsilon_L\epsilon_K = \epsilon_{K+L}$ is in $H \cap N$. If $K = T_0 + T_1, L = T_2 + T_3$ and $\Omega - (K + L) = T_4 + T_5$, T_i special 4-sets, a similar calculation shows that $\epsilon_{K+T_2+T_4}$ is in $H \cap N$. Since

$$\epsilon_{K+L}\epsilon_{K+T_2+T_4} = \epsilon_{T_3+T_4},$$

$H \cap N$ contains $\epsilon_{T_3+T_4}$. If M is any other 8-set in $R(C)$, and ξ sends $T_3 + T_4$ to M, then $\epsilon_M = \xi^{-1}\epsilon_{T_3+T_4}\xi$ is in $H \cap N$. Thus $H \cap N$ must contain all of N. Since N was maximal, $H = N$.

We now show that N cannot be normal in $\cdot 0$. Consider $\zeta_{T_0} = \eta_B\epsilon_{T_0}$ (Theorem 3.6 ff.) where

$$B = \{T_0, T_1, T_2, T_3, T_4, T_5\}$$

is a family of disjoint 4-sets, any two of which form an 8-set in $R(C)$. Recall that $\zeta = \zeta_{T_0}$, together with N, generate all of $\cdot 0$. Let θ in M_{24} send $T_1 + T_2$ to $T_3 + T_4$ and suppose that

$$T_i = \{4i, 4i + 1, 4i + 2, 4i + 3\}, \qquad i = 0, 1, 2, 3, 4$$

and

$$T_5 = \{20, 21, 22, \infty\}.$$

Suppose also that $(4)\theta = 12$, $(8)\theta = 13$. (The above distribution of the elements of Ω among the T_i is for convenience—see the material just preceding Figure 3.17 for the distribution that Conway used.) Assume that N is normal in $\cdot 0$. Then N contains $\zeta_{T_0}^{-1}\theta\zeta_{T_0} = \zeta_{T_0}\theta\zeta_{T_0}$ and

$$\begin{aligned}(4\mathbf{v}_4+4\mathbf{v}_8)\zeta_{T_0}\theta\zeta_{T_0} &= (4\mathbf{v}_4-2\mathbf{v}_{T_1}+4\mathbf{v}_8-2\mathbf{v}_{T_2})\theta\zeta_{T_0}\\ &= (4\mathbf{v}_{12}+4\mathbf{v}_{13}-2\mathbf{v}_{T_3+T_4})\zeta_{T_0}\\ &= 4\mathbf{v}_{12}-2\mathbf{v}_{T_3}+4\mathbf{v}_{13}-2\mathbf{v}_{T_3}+2\mathbf{v}_{T_3+T_4}\\ &= 4\mathbf{v}_{12}+4\mathbf{v}_{13}-2\mathbf{v}_{T_3}+2\mathbf{v}_{T_4}\\ &= 2\mathbf{v}_{12}+2\mathbf{v}_{13}-2\mathbf{v}_{14}-2\mathbf{v}_{15}+2\mathbf{v}_{16}\\ &\quad +2\mathbf{v}_{17}+2\mathbf{v}_{18}+2\mathbf{v}_{19}\end{aligned}$$

which is in Λ_2^2. This contradicts the fact that N preserves, as a whole, Λ_2^4. Thus H cannot exist, as postulated, and Theorem 3.18 is proved. ■

The two other new sporadic simple groups in $\cdot 0$, namely $\cdot 2$ and $\cdot 3$, are stabilizers of two points of Λ_2 whose differences are vectors of length $4\sqrt{2}$ and $4\sqrt{3}$ (i.e., the vectors are in Λ_2 and Λ_3), respectively [**25**, p. 241]. This completes the description of Conway's link.

Daniel Gorenstein, in an article on the classification of simple groups, has stated,

> ... if Conway had studied the Leech lattice some five years earlier, he would have discovered a total of seven new simple groups! [Besides the Mathieu groups, four other sporadic simple groups are contained in $\cdot 0$.] Unfortunately, he had to settle for three. However, as consolation, his paper on $\cdot 0$ will stand as one of the most elegant achievements of mathematics. [**48**, p. 106]

Conway's discoveries on a Saturday afternoon and evening end a colorful chapter in mathematics which began with the weekend encounter with a computer by Hamming over twenty-one years earlier. As we have shown, this route led, from a brief example given by Shannon, to the Golay codes. These inspired Leech's discovery of a new dense packing of spheres in E^{24}. And, in turn, the lattice of the sphere centers inspired Conway's research on simple groups.

However, Conway's work did not mark the end of investigation in either coding or sphere packing. Much effort is still being bent to the theory of coding [**14**], [**112**]. Moreover, other results in coding theory are being applied to sphere packing [**112**].

Furthermore, Conway's discoveries did not close the book on simple groups. An announcement (1979) of a conference on finite group theory optimistically stated,

> ... After more than a quarter of a century of intensive effort by several hundred people, the classification of all finite simple groups appears to be rapidly nearing conclusion. One can safely predict that by 1980 the general theory of simple groups will have been fully developed and that if the classification is not complete by then, all that will remain to analyze will be a very few specialized problems. [**94**]

Indeed, the work of Michael Aschbacher in 1980 essentially completed the classification [**38**].

The mathematics of the next few decades will evolve from the present as unpredictably, as serendipitously as the present has evolved from work stretching decades into the past. This golden age of mathematical discovery, of incredible sparks leaping from field to field, shows no sign of abatement.

DENSEST KNOWN SPHERE PACKINGS

The main feature of this Appendix is Table A1.1, a list of the densest known packings in E^n for $n = 1$ through 24, and for a few selected n larger than 24. The Table is essentially that in [**83**] updated by material from [**112**], [**113**] and [**95**]. It differs from that in [**83**], however, in that density, instead of center density is used. (Recall that the density, ρ, of a packing in E^n is the fraction of E^n within the spheres of the packing, while the center density, δ, is ρ divided by V_n, the volume of a sphere in the packing.) The latter is the number of centers of unit spheres per unit n-dimensional volume.

Several other comments on Table A1.1 are in order.

1. Under the column marked "Type," B indicates that both a lattice and a nonlattice packing with these parameters are known. L indicates that at present only a lattice packing is known, and N that only a nonlattice packing is known. A indicates a local arrangement of spheres touching one sphere.

2. N. J. A. Sloane kindly provided the sources [**112**], [**113**] and [**95**]. Specifically, the packing $P10c$ in E^{10} appears in [**113,** p. 31], the packings $P11c$ in E^{11} and one in E^{36} with no designation appear in [**112,** p. 118], and new bounds for contact numbers appear in [**95**].

3. All calculations converting from center density to density are given in truncated form and are obtained from Leech and Sloane's results, also given in truncated form.

4. The center density upper bound is Rogers', and was taken from [**74**]. For dimensions 32 and higher, it is Rogers' bound in the form

$$\log \rho \approx \tfrac{1}{2} n \log\left(\frac{n}{4e\pi}\right) + \left(\tfrac{3}{2}\right) \log n - \log \frac{e}{\sqrt{\pi}} + \frac{5\frac{1}{4} \log 2}{n + 2\frac{1}{2}},$$

the last term being approximate. For $8 < n \leq 32$, Leech used unpublished tables of the Schläfli functions $f_n(x)$ computed by G. R. Lang to calculate Rogers' bound as well as Coxeter's conjectured contact bound in E^{32} [**74**, p. 265].

5. The upper bound for the contact number in E^{32} is Coxeter's conjectured bound in [**30**]; all others are due to Odlyzko and Sloane in [**95**].

TABLE A1.1—

n	Name of packing	Type	Density Attained		Theoretical upper bound
1	$\Lambda_1 = A_1$	L	$2 \cdot \frac{1}{2}$	$= 1.0000$	1.0000
2	$\Lambda_2 = A_2$	L	$\pi \cdot \frac{1}{2\sqrt{3}}$	$= 0.9068$	0.9068
3	$\Lambda_3 = D_3$	B	$\frac{4\pi}{3} \cdot \frac{1}{4\sqrt{2}}$	$= 0.7404$	0.7796
4	$\Lambda_4 = D_4$	L	$\frac{\pi^2}{2} \cdot \frac{1}{8}$	$= 0.6168$	0.6477
5	$\Lambda_5 = D_5$	B	$\frac{8\pi^2}{15} \cdot \frac{1}{8\sqrt{2}}$	$= 0.4652$	0.5256
6	$\Lambda_6 = E_6$	B	$\frac{\pi^3}{6} \cdot \frac{1}{8\sqrt{3}}$	$= 0.3729$	0.4192
7	$\Lambda_7 = E_7$	B	$\frac{16\pi^3}{105} \cdot \frac{1}{16}$	$= 0.2952$	0.3298
8	$\Lambda_8 = E_8$	L	$\frac{\pi^4}{24} \cdot \frac{1}{16}$	$= 0.2536$	0.2567
9	$\Lambda_9 = T_9$	B	$\frac{32\pi^4}{945} \cdot \frac{1}{16\sqrt{2}}$	$= 0.1457$	0.1981
	$P9a$	N	$\frac{32\pi^4}{945} \cdot \frac{5}{128}$	$= 0.1288$	

—Continued

Density as percent of upper bound	Contact Number Maximum	Contact Number Average	Theoretical upper bound for maximum	Maximum as percent of upper bound	n
100.0	2	2	2	100.0	1
100.0	6	6	6	100.0	2
94.9	12	12	12	100.0	3
95.2	24	24	25	96.0	4
88.4	40	40	46	86.9	5
88.9	72	72	82	87.8	6
89.5	126	126	140	90.0	7
98.7	240	240	240	100.0	8
73.5	272	272	380	71.5	9
65.0	306	$235\frac{3}{5}$		80.5	

TABLE A1.1—

n	Name of packing	Type	Density Attained		Theoretical upper bound
10	$\Lambda_{10} = \Phi_{10}$	B	$\frac{\pi^5}{120}\cdot\frac{1}{16\sqrt{3}}$	$= 0.09202$	0.1518
	$P10a$	N	$\frac{\pi^5}{120}\cdot\frac{19}{512}$	$= 0.09463$	
	$P10b$	N	$\frac{\pi^5}{120}\cdot\frac{9}{256}$	$= 0.08965$	
	$P10c$	N	$\frac{\pi^5}{120}\cdot\frac{5}{128}$	$= 0.09961$	
11	K_{11}	L	$\frac{64\pi^5}{10395}\cdot\frac{1}{18\sqrt{3}}$	$= 0.06043$	0.1156
	$P11a$	N	$\frac{64\pi^5}{10395}\cdot\frac{9}{256}$	$= 0.06623$	
	$P11b$	A			
	$P11c$	N	$\frac{64\pi^5}{10395}\cdot\frac{53}{2048}$	$= 0.04875$	
12	K_{12}	L	$\frac{\pi^6}{720}\cdot\frac{1}{27}$	$= 0.04945$	0.08757
	$P12a$	N	$\frac{\pi^6}{720}\cdot\frac{9}{256}$	$= 0.04694$	
13	K_{13}	B	$\frac{128\pi^6}{135135}\cdot\frac{1}{18\sqrt{3}}$	$= 0.02920$	0.06604
	$P13a$	N	$\frac{128\pi^6}{135135}\cdot\frac{9}{256}$	$= 0.03201$	
14	Λ_{14}	B	$\frac{\pi^7}{5040}\cdot\frac{1}{16\sqrt{3}}$	$= 0.02162$	0.04960
	$P14b$	A			

—Continued

Density as percent of upper bound	Contact Number Maximum	Contact Number Average	Theoretical upper bound for maximum	Maximum as percent of upper bound	n
60.6	336	336	595	56.4	10
62.3	372	$353\frac{9}{19}$		62.5	
59.0	500	$340\frac{1}{3}$		84.0	
65.6	372	372		62.5	
52.2	432	432	915	47.2	11
57.2	566	$519\frac{7}{9}$		61.8	
	576			62.9	
42.1	582			63.6	
56.5	756	756	1416	53.3	12
53.6	840	$770\frac{2}{3}$		59.3	
44.2	918	918	2233	41.1	13
48.4	1130	$1060\frac{2}{3}$		50.6	
43.5	1422	1422	3492	40.7	14
	1582			45.3	

TABLE A1.1—

n	Name of packing	Type	Density Attained		Theoretical upper bound
15	Λ_{15}	B	$\frac{256\pi^7}{2027025}\cdot\frac{1}{16\sqrt{2}}$	$= 0.01685$	0.03713
	$P15a$	A			
16	Λ_{16}	B	$\frac{\pi^8}{40320}\cdot\frac{1}{16}$	$= 0.01470$	0.02770
17	Λ_{17}	B	$\frac{4^9 9!\pi^8}{18!}\cdot\frac{1}{16}$	$= 0.008811$	0.02061
18	$\Lambda_{18} = K_{18}$	B	$\frac{\pi^9}{9!}\cdot\frac{1}{8\sqrt{3}}$	$= 0.005928$	0.01530
19	Λ_{19}	B	$\frac{4^{10}10!\pi^9}{20!}\cdot\frac{1}{8\sqrt{2}}$	$= 0.004120$	0.01133
20	Λ_{20}	B	$\frac{\pi^{10}}{10!}\cdot\frac{1}{8}$	$= 0.003225$	0.008375
21	Λ_{21}	B	$\frac{4^{11}11!\pi^{10}}{22!}\cdot\frac{1}{4\sqrt{2}}$	$= 0.002465$	0.006177
22	Λ_{22}	L	$\frac{\pi^{11}}{11!}\cdot\frac{1}{2\sqrt{3}}$	$= 0.002127$	0.004549
23	Λ_{23}	L	$\frac{4^{12}12!\pi^{11}}{24!}\cdot\frac{1}{2}$	$= 0.001905$	0.003344
24	Λ_{24}	L	$\frac{\pi^{12}}{12!}\cdot 1$	$= 0.001929$	0.002455
32	Λ_{32}	L	$\frac{\pi^{16}}{16!}\cdot 1$	$= 4.303 \times 10^{-6}$	1.974×10^{-4}
36	Λ_{36}	L	$\frac{\pi^{18}}{18!}\cdot 2$	$= 2.775 \times 10^{-7}$	5.498×10^{-5}
		L	$\frac{\pi^{18}}{18!}\cdot 3.330$	$= 4.621 \times 10^{-7}$	

—Continued

Density as percent of upper bound	Contact Number Maximum	Contact Number Average	Theoretical upper bound for maximum	Maximum as percent of upper bound	n
45.3	2340	2340	5431	43.0	15
	2564			47.2	
53.0	4320	4320	8313	51.9	16
42.7	5346	5346	12215	43.7	17
38.7	7398	7398	17877	41.3	18
36.3	10668	10668	25901	41.1	19
38.5	17400	17400	37974	45.8	20
39.9	27720	27720	56852	48.7	21
46.7	49896	49896	86537	57.6	22
56.9	93150	93150	128096	72.7	23
78.5	196560	196560	196560	100.0	24
2.17	208320		6256830	3.32	32
0.504	234456				36
0.840					

TABLE A1.1—

n	Name of packing	Type	Density Attained		Theoretical upper bound
40	Λ_{40}	L	$\frac{\pi^{20}}{20!}\cdot 16$	$= 5.767 \times 10^{-8}$	1.518×10^{-5}
48	$P48p$	L	$\frac{\pi^{24}}{24!}\cdot 16832$	$= 2.317 \times 10^{-8}$	1.124×10^{-6}
	$P48q$	L	$\frac{\pi^{24}}{24!}\cdot 16832$	$= 2.317 \times 10^{-8}$	
60	$P60p$	L	$\frac{\pi^{30}}{30!}\cdot 95817$	$= 2.966 \times 10^{-13}$	2.160×10^{-8}
64	$P64a$	B	$\frac{\pi^{32}}{32!}\cdot 2^{16}$	$= 2.018 \times 10^{-15}$	5.743×10^{-9}
	$P64c$	N	$\frac{\pi^{32}}{32!}\cdot 2^{22}$	$= 1.292 \times 10^{-13}$	
128	$P128a$	B	$\frac{\pi^{64}}{64!}\cdot 2^{64}$	$= 9.552 \times 10^{-39}$	2.454×10^{-18}
	$P128b$	N	$\frac{\pi^{64}}{64!}\cdot 2^{85}$	$= 2.003 \times 10^{-32}$	
256	$P256a$	B	$\frac{\pi^{128}}{128!}\cdot 2^{192}$	$= 7.027 \times 10^{-95}$	2.715×10^{-37}
	$P256b$	N	$\frac{\pi^{128}}{128!}\cdot 2^{250}$	$= 2.025 \times 10^{-77}$	
512	$P512a$	B	$\frac{\pi^{256}}{256!}\cdot 2^{512}$	$= 2.912 \times 10^{-229}$	9.535×10^{-79}
	$P512b$	N	$\frac{\pi^{256}}{256!}\cdot 2^{698}$	$= 2.856 \times 10^{-173}$	

—Continued

Density as percent of upper bound	Contact Number Maximum	Contact Number Average	Theoretical upper bound for maximum	Maximum as percent of upper bound	n
0.379	531120				40
2.06	52416000				48
2.06	52416000				
0.00137	3908160				60
3.51×10^{-5}	9694080				64
2.24×10^{-3}					
3.89×10^{-19}					128
8.16×10^{-13}					
2.58×10^{-56}					256
7.45×10^{-39}					
3.05×10^{-149}					512
2.99×10^{-93}					

FURTHER PROPERTIES OF THE (24, 12) GOLAY CODE AND THE RELATED STEINER SYSTEM $S(5, 8, 24)$

These properties are needed at various places in the text. In particular, the number and weights of the various codewords are found in Table A2.2, and the number of distinct 8-sets which contain a given number of coordinates is found in Table A2.3.

For convenience, we shall use Leech's 12 by 24 matrix C (cf. Fig. 2.14) for generating the code. This is given again in Figure A2.1.

$$
\begin{bmatrix}
1&0&0&0&0&0&0&0&0&0&0&0&0&1&1&1&1&1&1&1&1&1&1&1\\
0&1&0&0&0&0&0&0&0&0&0&0&1&1&0&1&0&0&0&1&1&1&0&1\\
0&0&1&0&0&0&0&0&0&0&0&0&1&0&1&0&0&0&1&1&1&0&1&1\\
0&0&0&1&0&0&0&0&0&0&0&0&1&1&0&0&0&1&1&1&0&1&1&0\\
0&0&0&0&1&0&0&0&0&0&0&0&1&0&0&0&1&1&1&0&1&1&0&1\\
0&0&0&0&0&1&0&0&0&0&0&0&1&0&0&1&1&1&0&1&1&0&1&0\\
0&0&0&0&0&0&1&0&0&0&0&0&1&0&1&1&1&0&1&1&0&1&0&0\\
0&0&0&0&0&0&0&1&0&0&0&0&1&1&1&1&0&1&1&0&1&0&0&0\\
0&0&0&0&0&0&0&0&1&0&0&0&1&1&1&0&1&1&0&1&0&0&0&1\\
0&0&0&0&0&0&0&0&0&1&0&0&1&1&0&1&1&0&1&0&0&0&1&1\\
0&0&0&0&0&0&0&0&0&0&1&0&1&0&1&1&0&1&0&0&0&1&1&1\\
0&0&0&0&0&0&0&0&0&0&0&1&1&1&1&0&1&0&0&0&1&1&1&0
\end{bmatrix}
$$

FIG. A2.1. Leech's 12 by 24 matrix C.

LEMMA A2.1. *Any two vectors in $R(C)$ have an even number of 1's in common.*

A quick check shows the result true for any two rows of C. Let A and B be two vectors in $R(C)$. Denote the rows of $R(C)$ by $R_1, R_2, \ldots, R_{12}$. Then

$$A = \alpha_1 R_1 + \alpha_2 R_2 + \cdots + \alpha_{12} R_{12}$$

and

$$B = \beta_1 R_1 + \beta_2 R_2 + \cdots + \beta_{12} R_{12}$$

where the α_i's and β_j's are zeros and ones. Let AB be the *Boolean product* of A with B. That is, AB is the vector (not necessarily in $R(C)$) with 1's where both A and B have 1's, and 0's elsewhere. This is simply a componentwise multiplication. It is an exercise to show that this multiplication is distributive with respect to binary addition so that

$$\begin{aligned} AB &= (\alpha_1 R_1 + \cdots + \alpha_{12} R_{12})(\beta_1 R_1 + \cdots + \beta_{12} R_{12}) \\ &= \alpha_1 \beta_1 R_1 R_1 + \alpha_1 \beta_2 R_1 R_2 + \cdots + \alpha_1 \beta_{12} R_1 R_{12} \\ &\quad + \cdots \\ &\quad + \alpha_{12} \beta_1 R_{12} R_1 + \cdots + \alpha_{12} \beta_{12} R_{12} R_{12}. \end{aligned}$$

This calculation shows that AB is a (binary) sum of vectors each of which has an even number of 1's. Since the sum of any two vectors with an even number of 1's has an even number of 1's, so does AB and the result follows. ∎

COROLLARY A2.2. *Any vector in $R(C)$ has a multiple of four 1's.*

From Figure A2.1 it is clear that the result is true for any row of C. The number of 1's in the sum of any two rows of C is the sum of the number of 1's in each, minus twice the even number they have in common so that the result is a multiple of four. Since any vector in $R(C)$ is a sum of rows of C, the lemma, together with the above process, prove the corollary. ∎

COROLLARY A2.3. *Any vector in $R(C)$ has exactly* 0, 8, 12, 16 *or* 24 1*'s.*

As was shown by inspection (p. 90–94), any nontrivial vector in $R(C)$ has more than four 1's and, by complementation, fewer than twenty. By corollary A2.2, such vectors have 8, 12 or 16 1's and the result follows. ■

Since $R(C)$ has $2^{12} = 4096$ vectors and since there are exactly 759 vectors with eight 1's, it is possible to count the number of each type. Because the complement of an 8-set is a 16-set and these are the only 16-sets, there are 759 16-sets. Thus there are exactly $4096 - (2 \cdot 759 + 2) = 2576$ 12-sets. These results are collected in Table A2.2.

TABLE A2.2

Shape	Number
0^{24}	1
$0^{16}, 1^{8}$	759
$0^{12}, 1^{12}$	2,576
$0^{8}, 1^{16}$	759
1^{24}	1

The sets in $R(C)$.

LEMMA A2.4. *The twenty-four coordinates are evenly distributed among the* 8*-sets of $R(C)$.*

We will show that any element in $S = \{1, 2, 3, \ldots, 24\}$ appears in exactly 253 of the 759 8-sets. To do this, observe that any coordinate x_i is contained in $\binom{23}{4}$ 5-sets, since there are $\binom{23}{4}$ vertices of the unit 24-cube which have a 1 in the x_ith coordinate and a total of five 1's. Suppose x_i is in k 8-sets. For any 8-set in which x_i lies, x_i must also be in $\binom{7}{4}$ 5-sets contained in the 8-set. Since each 5-set is in exactly one 8-set, $\binom{23}{4} = k\binom{7}{4}$, so that $k = 253$. That is, 253 8-sets contain x_i. ■

Similarly, any pair of coordinate numbers, x_1, x_2, must be in exactly $\binom{22}{3}/\binom{6}{3} = 77$ 8-sets; any three, x_1, x_2 and x_3, must be in exactly $\binom{21}{2}/\binom{5}{2} = 21$ 8-sets, and any four, x_1, x_2, x_3 and x_4, must be in exactly $\binom{20}{1}/\binom{4}{1} = 5$ 8-sets. These results are collected in Table A2.3.

TABLE A2.3

Number of Coordinates	Number of Distinct 8-Sets
1	253
2	77
3	21
4	5
5	1

The number of distinct 8-sets containing a given number of coordinates.

In particular, given any four coordinates (a 4-set), we have

COROLLARY A2.5. *Any 4-set of S defines five other 4-sets, all pairwise disjoint, such that any two make up an 8-set.*

LEMMA A2.6. *Any 12-set of R(C) is the sum of two 8-sets of R(C).*

Pick any 5-set in the 12-set. By corollary A2.3, the corresponding 8-set has one additional element in common with the 12-set and the second 8-set is the sum of the 12-set and the first 8-set. It follows that the 12-set is the sum of the two 8-sets. ■

Note that for a given 12-set, the family of 8-sets corresponding to the 5-sets contained in the 12-set generate a family of 6-sets which are subsets of the 12-set. Furthermore, this family of 6-sets forms a Steiner system $S(5, 6, 12)$. Consequently, the $\binom{12}{5}/\binom{6}{5} = 132$ such 6-sets come in 66 disjoint pairs. This means that any 12-set of R(C) is the sum of two 8-sets in sixty-six ways.

COROLLARY A2.7. *Any element of $R(C)$ is a sum of* 8-*sets of $R(C)$.*

By the lemma, the first row of C (cf. Figure A2.1) is the sum of two 8-sets and the result follows. ■

APPENDIX **3**

A CALCULATION OF THE NUMBER OF SPHERES WITH CENTERS IN Λ_2 ADJACENT TO ONE, TWO, THREE AND FOUR ADJACENT SPHERES WITH CENTERS IN Λ_2.

The first part of the calculation will repeat that found on pages 111–112. For ease of calculation, we start with a vector, $\mathbf{v}_1$, of shape $(0^{22}, 4^2)$ and assume that the 4's are in the first two coordinates. Vectors in Λ_2^4 at minimum distance from $\mathbf{v}_1$ must have one 4 in one of the first two coordinates and one ± 4 in the last twenty-two coordinates. (Recall from Chapter 3 that Λ_2^2 is the set of vectors in Λ of shape $(0^{16}, \pm 2^8)$, Λ_2^3, those of shape $(\pm 1^{23}, \pm 3)$ and Λ_2^4, those of shape $(0^{22}, \pm 4^2)$.) The total number of such vectors is $2 \cdot 2 \cdot 22 = 88$. Vectors in Λ_2^2 at minimum distance from $\mathbf{v}_1$ have 2's in the first two coordinates and ± 2's in six other positions. Since 77 8-sets contain the first two coordinates and since each vector in Λ_2^2 has an even number of 2's and -2's, there are five free sign choices for the six remaining positions in the 8-set which yields a total of $77 \cdot 2^5 = 2464$. Finally, those in Λ_2^3 closest to $\mathbf{v}_1$ will have one 1 and one 3 in the first two coordinates. The paragraph preceding Table 2.21 shows that the first two positions of the corresponding vectors in $R(C)$ must differ (one will contain a 0, the other a 1). Since exactly half the vectors of $R(C)$ do, the number of vertices in Λ_2^3 adjacent to $\mathbf{v}_1$ is $2^{11} = 2048$. All told, there are $88 + 2464 + 2048 = 4600$ vectors in Λ_2 adjacent to $\mathbf{v}_1$.

Since the vector $\mathbf{v}_2$, with a 4 in the first and third coordinates, is adjacent to $\mathbf{v}_1$, we now seek those vectors in Λ_2 adjacent to both $\mathbf{v}_1$ and $\mathbf{v}_2$:

$$\mathbf{v}_1 = (4, 4, 0, 0, 0, 0, 0, 0, \ldots, 0)$$
$$\mathbf{v}_2 = (4, 0, 4, 0, 0, 0, 0, 0, \ldots, 0).$$

Those in Λ_2^4 have either one 4 in the first position and a ± 4 in one of the last twenty-one positions, or one 4 in the second and third coordinates. The total number is $2 \cdot 21 + 1 = 43$. Those vectors in Λ_2^2 closest to $\mathbf{v}_1$ and $\mathbf{v}_2$ have 2's in the first three coordinates. The other five ± 2's must be in the last twenty-one positions. Since the first three coordinates are in 21 8-sets (cf. Table A2.3) and since there must be an even number of 2's and -2's, there are four free sign choices for the five remaining positions in each 8-set which yields a total of $21 \cdot 2^4 = 336$. Lastly, those in Λ_2^3 must have a 3 in the first coordinate and 1's in the next two. We already know that half the vectors of $R(C)$ differ in the first two coordinates. Half of these will have a 1 in the first coordinate and a 0 in the second and half again will have a 0 in the third coordinate yielding $2^{11} \cdot \frac{1}{2} \cdot \frac{1}{2} = 512$ of this type. The total is thus $43 + 336 + 512 = 891$.

We now add a third vector $\mathbf{v}_3$:

$$\mathbf{v}_1 = (4, 4, 0, 0, 0, 0, 0, 0, \ldots, 0)$$
$$\mathbf{v}_2 = (4, 0, 4, 0, 0, 0, 0, 0, \ldots, 0)$$
$$\mathbf{v}_3 = (0, 4, 4, 0, 0, 0, 0, 0, \ldots, 0).$$

This time there are no closest vectors from either Λ_2^4 or Λ_2^3. Those from Λ_2^2 are as in the last case and number 336. Thus there are 336 vertices in Λ_2 at minimum distance from the equilateral triangle with vertices $\mathbf{v}_1$, $\mathbf{v}_2$ and $\mathbf{v}_3$.

Let $\mathbf{v}_4$ be the vector of shape $(0^{16}, 2^8)$, the 2's being in the first eight coordinates; this we may assume after making a suitable interchange of coordinates:

$$\begin{aligned}
\mathbf{v}_1 &= (4, 4, 0, 0, 0, 0, 0, 0, 0, \ldots, 0)\\
\mathbf{v}_2 &= (4, 0, 4, 0, 0, 0, 0, 0, 0, \ldots, 0)\\
\mathbf{v}_3 &= (0, 4, 4, 0, 0, 0, 0, 0, 0, \ldots, 0)\\
\mathbf{v}_4 &= (2, 2, 2, 2, 2, 2, 2, 2, 0, \ldots, 0).
\end{aligned}$$

Any vector at minimum distance from all of these points must be, as in the last case, from Λ_2^2. There are two types. The first has six 2's, three of which are in the first three coordinates. The other three, together with two -2's, are in the other five positions of the 2's in $\mathbf{v}_4$. The total number of vectors of this type is $\binom{5}{2} = 10$. The second type matches the 2's of $\mathbf{v}_4$ in exactly four places, three of which are the first three coordinates. These positions all have a 2. The other four positions contain an even number of 2's and -2's and form one of the four possible 4-sets disjoint from the two 4-sets making up the first eight positions of $\mathbf{v}_4$ (cf. Corollary A2.5). With three free sign choices for the last four positions, the number of this type is thus $5 \cdot 4 \cdot 2^3 = 160$. This makes a total of $10 + 160 = 170$ elements of Λ_2 adjacent to the four vertices $\mathbf{v}_1$, $\mathbf{v}_2$, $\mathbf{v}_3$ and $\mathbf{v}_4$.

Let us look a little closer at the set of 10 points, which are listed in Figure A3.1:

$$\begin{aligned}
&(2, 2, 2, -2, -2, 2, 2, 2, 0, \ldots, 0)\\
&(2, 2, 2, -2, 2, -2, 2, 2, 0, \ldots, 0)\\
&(2, 2, 2, -2, 2, 2, -2, 2, 0, \ldots, 0)\\
&(2, 2, 2, -2, 2, 2, 2, -2, 0, \ldots, 0)\\
&(2, 2, 2, 2, -2, -2, 2, 2, 0, \ldots, 0)\\
&(2, 2, 2, 2, -2, 2, -2, 2, 0, \ldots, 0)\\
&(2, 2, 2, 2, -2, 2, 2, -2, 0, \ldots, 0)\\
&(2, 2, 2, 2, 2, -2, -2, 2, 0, \ldots, 0)\\
&(2, 2, 2, 2, 2, -2, 2, -2, 0, \ldots, 0)\\
&(2, 2, 2, 2, 2, 2, -2, -2, 0, \ldots, 0)
\end{aligned}$$

FIG. A3.1

They differ in only five coordinates, the fourth through the eighth, and hence lie in a four-dimensional subspace of E^{24}. Leech noticed, however, that these are the midpoints of the line segments joining the following five vertices in E^{24}:

$$\begin{array}{l}(2, 2, 2, -6, 2, 2, 2, 2, 0, \ldots, 0)\\(2, 2, 2, 2, -6, 2, 2, 2, 0, \ldots, 0)\\(2, 2, 2, 2, 2, -6, 2, 2, 0, \ldots, 0)\\(2, 2, 2, 2, 2, 2, -6, 2, 0, \ldots, 0)\\(2, 2, 2, 2, 2, 2, 2, -6, 0, \ldots, 0),\end{array}$$

which are the vertices of a 4-dimensional regular simplex [**80**]. Such a simplex has symmetry group S_5 of order 120, which also permutes the 10 midpoints. This observation shows why he expected the number 120 to be a factor of one of the two numbers he conjectured as divisors of the order of the symmetry group of P, the polytope whose vertices are the vectors of Λ_2.

All these calculations can be duplicated for other sets of four vectors each at minimum distance from the other three. However, it is not as easy to see that the 170 vectors closest to four vertices of a regular tetrahedron break down into 10 + 160 points, the ten of which are inequivalent to the other 160.

THE MATHIEU GROUP M_{24} AND THE ORDER OF M_{22}

The major portion of this Appendix is devoted to the completion of the proof that M_{24}, as defined by Conway (Figure 3.9 ff.) is the automorphism group of a Steiner system $S(5, 8, 24)$ (Corollary A4.5). The Appendix concludes with a derivation of the order of M_{22}.

DEFINITION: M_{24} is the subgroup of S_{24} generated by:

$$\begin{aligned}\alpha &= (\infty)(0\ 1\ 2\ \ldots\ 22)\\ \gamma &= (0\ \infty)(1\ 22)(2\ 11)(3\ 15)(4\ 17)(5\ 9)(6\ 19)\\ &\quad (7\ 13)(8\ 20)(10\ 16)(12\ 21)(14\ 18)\\ \delta &= (\infty)(0)(3)(15)(1\ 18\ 4\ 2\ 6)(5\ 21\ 20\ 10\ 7)\\ &\quad (8\ 16\ 13\ 9\ 12)(11\ 19\ 22\ 14\ 17).\end{aligned}$$

We have already shown how this group arises from $PSL_2(23)$ and that it is quintuply transitive. The rest of this Appendix follows the treatment of M_{24} by Conway in [**25**].

THEOREM A4.1. *M_{24} preserves a Steiner system $S(5, 8, 24)$.*

This theorem will be proved in three steps. In the first, the set $P(\Omega)$ of all subsets of $\Omega = \{0, 1, 2, \ldots, 22, \infty\}$ is turned into a 24-dimensional vector space over the field GF(2) by defining the sum $A + B$ of two subsets of Ω to be their symmetric difference, and a certain 12-dimensional subspace G is singled out.

The second step will show that M_{24} preserves G, and finally, the last step will show that G has a subset which is a Steiner system $S(5, 8, 24)$.

The first part of the first step is easy if each subset of Ω is thought of as a 24-tuple with 1's in the coordinate positions of the elements of the subset and 0's elsewhere. Addition by symmetric difference then becomes vector addition modulo 2. So $P(\Omega)$ is now a vector space V_{24}.

Define G as the subspace of V_{24} spanned by the twenty-three vectors $Q(i) = (Q)\alpha^i, i = 0, 1, 2, \ldots, 22$ and $Q(\infty) = \Omega - Q$. (See p. 134 for the definition of $Q = \{0, 1, 2, 3, 4, 6, 8, 9, 12, 13, 16, 18\}$.) Note that $Q(0) = Q$. For convenience, these twenty-four vectors are listed in coordinate form in Figure A4.1.

$Q(0)$	1	1	1	1	1	0	1	0	1	1	0	0	1	1	0	0	1	0	1	0	0	0	0	0
$Q(1)$	0	1	1	1	1	1	0	1	0	1	1	0	0	1	1	0	0	1	0	1	0	0	0	0
$Q(2)$	0	0	1	1	1	1	1	0	1	0	1	1	0	0	1	1	0	0	1	0	1	0	0	0
$Q(3)$	0	0	0	1	1	1	1	1	0	1	0	1	1	0	0	1	1	0	0	1	0	1	0	0
$Q(4)$	0	0	0	0	1	1	1	1	1	0	1	0	1	1	0	0	1	1	0	0	1	0	1	0
$Q(5)$	1	0	0	0	0	1	1	1	1	1	0	1	0	1	1	0	0	1	1	0	0	1	0	0
$Q(6)$	0	1	0	0	0	0	1	1	1	1	1	0	1	0	1	1	0	0	1	1	0	0	1	0
$Q(7)$	1	0	1	0	0	0	0	1	1	1	1	1	0	1	0	1	1	0	0	1	1	0	0	0
$Q(8)$	0	1	0	1	0	0	0	0	1	1	1	1	1	0	1	0	1	1	0	0	1	1	0	0
$Q(9)$	0	0	1	0	1	0	0	0	0	1	1	1	1	1	0	1	0	1	1	0	0	1	1	0
$Q(10)$	1	0	0	1	0	1	0	0	0	0	1	1	1	1	1	0	1	0	1	1	0	0	1	0
$Q(11)$	1	1	0	0	1	0	1	0	0	0	0	1	1	1	1	1	0	1	0	1	1	0	0	0
$Q(12)$	0	1	1	0	0	1	0	1	0	0	0	0	1	1	1	1	1	0	1	0	1	1	0	0
$Q(13)$	0	0	1	1	0	0	1	0	1	0	0	0	0	1	1	1	1	1	0	1	0	1	1	0
$Q(14)$	1	0	0	1	1	0	0	1	0	1	0	0	0	0	1	1	1	1	1	0	1	0	1	0
$Q(15)$	1	1	0	0	1	1	0	0	1	0	1	0	0	0	0	1	1	1	1	1	0	1	0	0
$Q(16)$	0	1	1	0	0	1	1	0	0	1	0	1	0	0	0	0	1	1	1	1	1	0	1	0
$Q(17)$	1	0	1	1	0	0	1	1	0	0	1	0	1	0	0	0	0	1	1	1	1	1	0	0
$Q(18)$	0	1	0	1	1	0	0	1	1	0	0	1	0	1	0	0	0	0	1	1	1	1	1	0
$Q(19)$	1	0	1	0	1	1	0	0	1	1	0	0	1	0	1	0	0	0	0	1	1	1	1	0
$Q(20)$	1	1	0	1	0	1	1	0	0	1	1	0	0	1	0	1	0	0	0	0	1	1	1	0
$Q(21)$	1	1	1	0	1	0	1	1	0	0	1	1	0	0	1	0	1	0	0	0	0	1	1	0
$Q(22)$	1	1	1	1	0	1	0	1	1	0	0	1	1	0	0	1	0	1	0	0	0	0	1	0
$Q(\infty)$	0	0	0	0	0	1	0	1	0	0	1	1	0	0	1	1	0	1	0	1	1	1	1	1

FIG. A4.1. The vectors generating $G \subset V_{24}$.

For any nonempty subset S of Ω, denote the sum $\Sigma Q_i, i \in S$, by $Q(S)$. With the aid of Table A4.1, it can be shown that

$$\begin{aligned}
&Q(0, 2, 20, 21) &&= \{0, 1, 2, 3, 4, 7, 10, 12\},\\
&Q(1, 3, 21, 22) &&= \{1, 2, 3, 4, 5, 8, 11, 13\},\\
&Q(2, 4, 22, 0) &&= \{2, 3, 4, 5, 6, 9, 12, 14\},\\
&\quad\vdots\\
&Q(i, i+2, i+20, i+21) &&= \{i, i+1, i+2, i+3, i+4, i+7, i+10, i+12\},\\
&\quad\vdots\\
\text{and } &Q(10, 12, 7, 8) &&= \{10, 11, 12, 13, 14, 17, 20, 22\}.
\end{aligned}$$

The coordinate form of these vectors is given below:

```
1 1 1 1 1 0 0 1 0 0 1 0 1 0 0 0 0 0 0 0 0 0 0 0
0 1 1 1 1 1 0 0 1 0 0 1 0 1 0 0 0 0 0 0 0 0 0 0
0 0 1 1 1 1 1 0 0 1 0 0 1 0 1 0 0 0 0 0 0 0 0 0
0 0 0 1 1 1 1 1 0 0 1 0 0 1 0 1 0 0 0 0 0 0 0 0
0 0 0 0 1 1 1 1 1 0 0 1 0 0 1 0 1 0 0 0 0 0 0 0
0 0 0 0 0 1 1 1 1 1 0 0 1 0 0 1 0 1 0 0 0 0 0 0
0 0 0 0 0 0 1 1 1 1 1 0 0 1 0 0 1 0 1 0 0 0 0 0
0 0 0 0 0 0 0 1 1 1 1 1 0 0 1 0 0 1 0 1 0 0 0 0
0 0 0 0 0 0 0 0 1 1 1 1 1 0 0 1 0 0 1 0 1 0 0 0
0 0 0 0 0 0 0 0 0 1 1 1 1 1 0 0 1 0 0 1 0 1 0 0
0 0 0 0 0 0 0 0 0 0 1 1 1 1 1 0 0 1 0 0 1 0 1 0
```

These eleven vectors, together with $Q(\infty)$, form a linearly independent set. Thus G has dimension at least twelve.

We will show that the dimension of G is twelve. Suppose thirteen of the $Q(i)$ are linearly independent. Writing them in coordinate form, we construct the 13 by 24 matrix with the thirteen independent $Q(i)$ as rows. Again, Table A4.1 shows that at least one row, say row one of this matrix, has a 1 in the first column. Replace the other rows that have a 1 in the first column by their sum with the first row. The resulting matrix then has a 1 in the first entry of the first column while the rest of that column has only 0's. The new matrix still has thirteen linearly independent rows. Pick the first column, to the right of column one, that has a nonzero entry not in the first row. Eliminate the rest of the 1's in that column in a similar way. Continue this process. Upon rearrangement of the columns, we get a matrix N of the form:

$$\begin{bmatrix} 1 & 0 & 0 & \cdots 0 & a_{1,14} & a_{1,15} & \cdots & a_{1,24} \\ 0 & 1 & 0 & \cdot & \cdot & & & \cdot \\ 0 & 0 & 1 & \cdot & \cdot & & & \cdot \\ \cdot & & & \cdot & \cdot & & & \cdot \\ \cdot & & & \cdot & \cdot & & & \cdot \\ \cdot & & & \cdot & \cdot & & & \cdot \\ 0 & 0 & 0 & \cdots 1 & a_{13,14} & a_{13,15} & \cdots & a_{13,24} \end{bmatrix}$$

The matrix N.

Since each $Q(i)$ has an even number of 1's, the same must be true of the rows of N. This also shows that the rows of N are pairwise orthogonal. Denote the last eleven entries in each row, as an 11-tuple, by r_i, $i = 1, 2, 3, \ldots, 13$. It is not hard to see that the dot product (r_i, r_j) equals 0 if $i \neq j$ and that $(r_i, r_i) = 1$ since each r_i has an odd number of 1's. This implies that the r_i, as vectors, are linearly independent, otherwise (relabeling if necessary) the equation

$$r_1 + r_2 + \cdots + r_i = 0$$

would imply that

$$1 = (r_1, r_1 + r_2 + \cdots + r_i) = 0.$$

But the r_i cannot be linearly independent since they belong to an 11-dimensional vector space. Thus the dimension of G is twelve.

To show that M_{24} takes G into itself, we will consider the effect of M_{24} on the generators of G.

First of all, α simply permutes the $Q(i)$.

A short computation shows that

$$\gamma\delta^2 = (\infty\ 0)(1\ 17\ 6\ 14\ 2\ 22\ 4\ 19\ 18\ 11)$$
$$(3\ 15)(5\ 8\ 7\ 12\ 10\ 9\ 20\ 13\ 21\ 16)$$

and that $(\gamma\delta^2)^5 = \gamma$ while $(\gamma\delta^2)^8 = \delta$. Thus M_{24} is generated by α and $\gamma\delta^2$. So all that is left is to check the action of $\gamma\delta^2$ on the $Q(i)$.

It is easy to see that $(Q(0))\gamma\delta^2 = Q(\infty)$ and that $(Q(\infty))\gamma\delta^2 = Q(0)$. A computation using Table A4.1 shows further that

$$(Q(1))\gamma\delta^2 = Q(2, 11, 20)$$

while

$$(Q(22))\gamma\delta^2 = Q(1, 20, 22, \infty).$$

Next, recall that

$$\beta = (\infty)(0)(1\ 2\ 4\ 8\ 16\ 9\ 18\ 13\ 3\ 6\ 12)$$
$$(5\ 10\ 20\ 17\ 11\ 22\ 21\ 19\ 15\ 7\ 14).$$

Another computation shows that $\beta\gamma\delta^2 = \gamma\delta^2\beta^2$. We use this fact to calculate $(Q(i))\gamma\delta^2$ for the other values of i. Note first that $(Q(i))\beta = Q(2i)$. Thus

$$(Q(2i))\gamma\delta^2 = (Q(i))\beta\gamma\delta^2 = (Q(i))\gamma\delta^2\beta^2$$

is in G. If i is in Q, a series of these operations reduces $(Q(i))\gamma\delta^2$ to $(Q(1))\gamma\delta^2\beta^{2l}$, which is in G. A similar result holds if i is in $\Omega - Q$. Thus G is preserved by M_{24}.

Since $\{0, 1, 2, 3, 4, 7, 10, 12\}$ is an element of G, the quintuple transitivity of M_{24} and the fact that G is preserved by M_{24} imply that any 5-set of Ω is contained in at least one 8-element set of G. If G has a set with five, six or seven elements, this set can be sent by M_{24} to one intersecting the above 8-element set in at least five elements. The sum of these two sets is then an element of G with at least one and at most five elements. By taking appropriate sums with the above 8-element set, we conclude that G contains a 2-set and thus all subsets of Ω with an even number of elements. This is impossible since G has only $2^{12} = 4096$ elements.

For example, suppose $\{0, 1, 2, 3, 4, 5\}$ is an element of G. Then

$$\{0, 1, 2, 3, 4, 5\} + \{0, 1, 2, 3, 4, 7, 10, 12\} = \{5, 7, 10, 12\}.$$

This set can be sent to $\{5, 7, 10, 11\}$ and their sum is the 2-set $\{11, 12\}$. Thus all 2-sets and hence all even-order sets of Ω would be in G. This impossibility shows that the smallest nonempty sets in G are 8-element sets. If two such sets were to contain the same 5-set, their sum would be a set in G with at most six elements, which is impossible. This shows that each 5-set of Ω is contained in exactly one 8-element set of G. In other words, the family of 8-element sets of G form a Steiner system $S(5, 8, 24)$. This concludes the proof of Theorem A4.1. ■

We are now in a position to compute the order of M_{24}.

THEOREM A4.2. $|M_{24}| = 244{,}823{,}040$.

For convenience, the first part of the proof will be extracted as a lemma. Let G be a permutation group on an n-set T. Let W be an r-subset of T. The *pointwise stabilizer* of W is the subgroup of G which not only fixes W as a whole, but fixes each point of W as well.

LEMMA A4.3. *The pointwise stabilizer of an* 8-*set is transitive on the remaining sixteen points of* Ω.

Note first that the permutation

$$\alpha^5\delta = (\infty)(0\ 21\ 3\ 16\ 20\ 6\ 19\ 18)(1)(2\ 5\ 7\ 8\ 9\ 17\ 14\ 22)(4\ 12\ 11\ 13)(10\ 15)$$

has structure $1^2 2^1 4^1 8^2$. Thus $(\alpha^5\delta)^4$ has structure $1^8 2^8$. Let the eight 1-cycles be $(a_1)(a_2), \ldots, (a_8)$ and suppose $K =$

$\{a_1, a_2, \ldots, a_8\}$ is not an 8-set. Let A be the 8-set containing a_1, a_2, a_3, a_4 and a_5. Since $(\alpha^5\delta)^4$ must fix A, A contains an additional point of K, say a_6, and the points of some 2-cycle, say a_9 and a_{10}. Let B be the 8-set containing a_3, a_4, a_5, a_6 and a_7. Then B must contain a_8 and the points of some other 2-cycle, say a_{11} and a_{12}. Let D be the 8-set containing a_1, a_3, a_4, a_5 and a_8. See Figure A4.2. Then D must contain an additional element of K. This element cannot be either a_2 or a_6 because, in either case, A and D would have five elements in common. Similarly, it cannot be a_7. This contradicts the fact that K was supposed not to be an 8-set. Thus K is an 8-set.

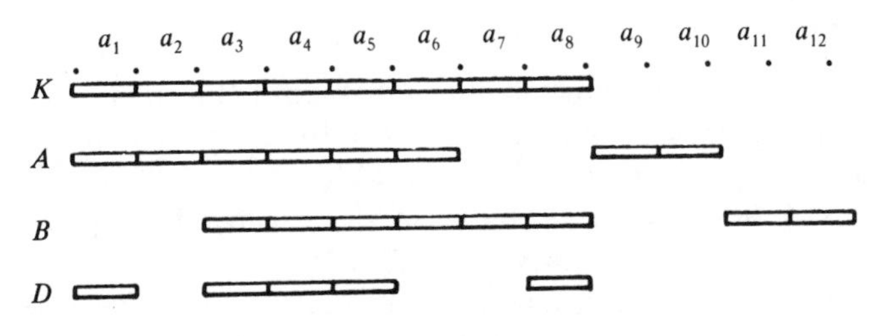

FIG. A4.2

Since M_{24} is quintuply transitive, the appropriate choice of ϕ yields a permutation with the same structure as $(\alpha^5\delta)^4$ but where the 1-cycles form any given 8-set. Consider the permutation $\phi^{-1}(\alpha^5\delta)^4\phi$.

Denote the pointwise stabilizer of an 8-set K by P_K. Clearly, P_K is a subgroup of M_{24} which contains a permutation with structure $1^8\,2^8$. We now show that P_K is transitive on $\Omega - K$. Designate the permutation of P_K with structure $1^8\,2^8$ by η.

The permutation $\delta\alpha^{11}$ equals

$$(\infty)(0\ 11\ 7\ 16\ 1\ 6\ 12\ 19\ 10\ 18\ 15\ 3\ 14\ 5\ 9)$$
$$(2\ 17\ 22)(4\ 13\ 20\ 21\ 8)$$

and has structure $1^1\,3^1\,5^1\,15^1$. Thus M_{24} has a permutation, θ, with this structure where the points of the 5-cycle are in K. Since θ must fix K as a whole, the 3-cycle of θ must contain the remaining points of K.

Consider the permutations $\theta^{-1}\eta\theta$, $(\theta^2)^{-1}\eta\theta^2, \ldots,$ $(\theta^{14})^{-1}\eta\theta^{14}$. These are all in P_K and have structure $1^8\,2^8$. If the 1-cycle of θ is (a_9) and if the 2-cycle of η containing a_9 is $(a_9 a_{10})$, each of the above permutations contains exactly one of the following 2-cycles: $(a_9\, a_{11})$, $(a_9\, a_{12})$, $(a_9\, a_{13})$, $\ldots$, $(a_9\, a_{24})$. This holds since a_{10} must be in the 15-cycle of θ.

It is now easy to see that P_K has at least sixteen elements, fifteen being of order two, and that P_K is transitive on $\Omega - K$. This proves the lemma. ■

The permutation $(\delta a^{11})^5$ has structure $1^6\,3^6$, which is of order three. This shows that the subgroup of M_{24}, fixing a 5-set of Ω pointwise, has order at least $3 \cdot 16$ since it contains P_K as a subgroup and $(\delta\alpha^{11})^5$ is not in P_K.

We now show that the index, in M_{24}, of any subgroup fixing a 5-set of Ω pointwise is $24\cdot 23\cdot 22\cdot 21\cdot 20$. Since M_{24} is quintuply transitive, any ordered 5-set, D, can be sent to $24\cdot 23\cdot 22\cdot 21\cdot 20$ distinct ordered 5-sets. Denote the corresponding permutations by $\theta_1, \theta_2, \theta_3, \ldots$. Denote the subgroup fixing D pointwise by R. If θ, in M_{24}, sends D to the same ordered 5-set as does θ_i, then $\theta\theta_i^{-1}$ is in R, so that θ is in the right coset $R\theta_i$. Thus, the $R\theta_i$ fill out M_{24}. If $i \neq j$, it is clear that $R\theta_i \neq R\theta_j$. Thus the order of M_{24} is at least

$$24\cdot 23\cdot 22\cdot 21\cdot 20\cdot 48 = 244{,}823{,}040.$$

Let F be the subgroup of M_{24} which carries the 8-set $\{a_1, a_2, \ldots, a_8\}$ into itself. Let H be the subgroup of F which, in addition, fixes the element of a_9 of Ω. Denote the fifteen elements of P_K of order two by $\eta_1, \eta_2, \eta_3, \ldots, \eta_{15}$ and consider the cosets $H, H\eta_1, H\eta_2, \ldots, H\eta_{15}$ of H in F. These are distinct since the η_i's must move a_9 to distinct elements of Ω —

K. If some element η of F sends a_9 to the same place as η_i (which must happen for some i if that element does not fix a_9), $\eta\eta_i^{-1}$ is in H so that η is in $H\eta_i$. This shows that H has index sixteen in F.

We now show that H is isomorphic to A_8, the alternating group on eight letters. Since the Steiner systems $S(5, 8, 24)$ are all isomorphic to the 8-sets of $R(C)$, the row space of the matrix C of Chapter 2, the same properties hold for G as for $R(C)$. Thus, of the thirty 8-sets disjoint from K, fifteen are disjoint from $\{a_1, a_2, \ldots, a_9\}$. These, together with $\emptyset$, form a subspace V of G of dimension four. H acts naturally on V, that is, any element of H fixes $\{a_1, \ldots, a_8\}$ as a whole, fixes a_9, and permutes the elements of V.

Suppose an element η of H acts trivially on V. We will show that this element is the identity of M_{24}. Let T_1, T_2, T_3 and T_4 be a family of disjoint 4-sets which, pairwise, form six 8-sets in $\Omega - K$. Three of the 8-sets are in V and are schematically represented in Figure A4.3 by $\mathbf{v}_1$, $\mathbf{v}_2$ and $\mathbf{v}_3$. (All computations will be made with the aid of this figure.)

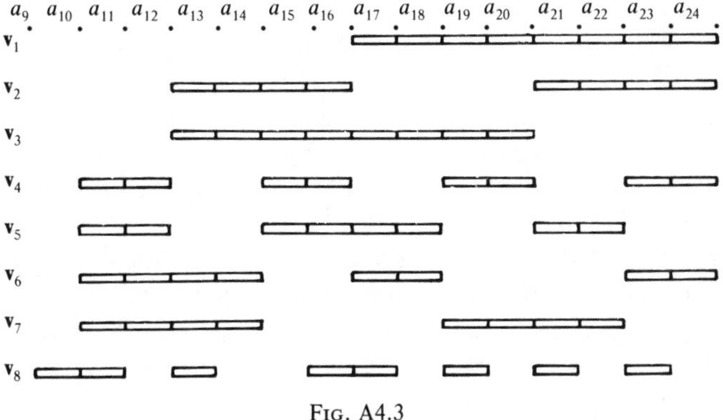

FIG. A4.3

The vectors in V must have (pairwise) four elements in common. This rules out the possibility of any such vector intersecting any of the three 4-sets in one or three elements. This shows that η fixes a_9, a_{10} and each of the pairs $\{a_{11}, a_{12}\}$, $\{a_{13}, a_{14}\}$, $\ldots$, $\{a_{23}, a_{24}\}$. The vector $\mathbf{v}_4$, together with $\mathbf{v}_1$, $\mathbf{v}_2$ and $\mathbf{v}_3$, generate eight vectors which still do not exhaust V. Thus, there is an additional vector, $\mathbf{v}_8$, which will contain exactly one element from each of the 2-blocks that form $\mathbf{v}_7$. This shows that η must fix $\Omega - K$ pointwise. If η moves a point of K, say $(a_1)\eta = a_5$, then the 8-set containing a_1, a_2, a_3, a_4 and a_9, say

$$\{a_1, a_2, a_3, a_4, a_9, a_{10}, a_{11}, a_{12}\}$$

would be sent to

$$\{a_5, a_6, a_7, a_8, a_9, a_{10}, a_{11}, a_{12}\};$$

thus η moves every element in K. Any 8-set intersecting K in exactly $\{a_1, a_2\}$ must be sent to itself by η. But then an 8-set intersecting K in exactly $\{a_2, a_3\}$ and its distinct image under η have six elements in common, which is impossible.

Thus η must fix K pointwise, that is, it is the identity since it already fixes $\Omega - K$ pointwise. This means that nontrivial elements of H act nontrivially on V so that H is (isomorphic to) a subgroup of the automorphism group $\mathrm{GL}_4(2)$ of V, the set of 4 by 4 matrices over GF(2) with nonzero determinant. We now count the order of this group.

The first row may be chosen in $2^4 - 1$ ways since only the all-zero row is excluded. Once the first row is chosen, the second row may be chosen in $2^4 - 2$ ways since it can be neither the all-zero row nor the first row. Since the third row cannot equal either of the first two rows, their sum or the all-zero row, it can be chosen in at most $2^4 - 4$ ways. Similarly, the last row can be chosen in at most $2^4 - 8$ ways. Thus, the order of the group $\mathrm{GL}_4(2)$ is

$$(16 - 1)(16 - 2)(16 - 4)(16 - 8) = 20160.$$

Each element of H induces a permutation of the set $\{a_1, a_2, \ldots, a_8\}$. In particular, for each 5-subset of $K = \{a_1, a_2, \ldots, a_8\}$, M_{24} has a permutation with structure $1^1\,3^1\,5^1\,15^1$ where both the 5-cycle and the 3-cycle permute elements in K. By conjugating with a permutation of structure $1^8\,2^8$, we can choose the 1-cycle of the permutation with structure $1^1\,3^1\,5^1\,15^1$ to be a_9 (by Lemma A4.3). Thus, for any 3-set of K, H contains a permutation that has that 3-set as a 3-cycle. This means that H has at least as many permutations as those generated by the 3-cycles on eight letters.

It is well known that the set of 3-cycles generate A_8 (see [**16,** p. 155]). Thus A_8 is a subgroup of H. But $|A_8| = 8!/2 = 20160$. This means that $A_8 \cong H \cong \mathrm{GL}_4(2)$, and thus $|H| = 20160$.

Since $[M_{24}:F] = 759$ and $[F:H] = 16$, $|M_{24}| = 759 \cdot 16 \cdot 20160 = 244{,}823{,}040$. This completes the proof of the theorem. ■

COROLLARY A4.4. $|R| = 48$, $|P_K| = 16$.

COROLLARY A4.5. *M_{24} is isomorphic to the automorphism group of a Steiner system $S(5, 8, 24)$.*

Let the automorphism group of a Steiner system $S(5, 8, 24)$ be denoted by M. By Theorem A4.1, M contains M_{24}. However, the proof of Theorem A4.2 goes through unchanged for M. Thus $|M| = |M_{24}|$ and $M = M_{24}$. ■

LEMMA A4.6. *Let G be a transitive permutation group on n letters and H a subgroup of G fixing one of the letters. Then $[G:H] = n$.*

Suppose G is transitive on $\{x_1, \ldots, x_n\}$ and H fixes x_1. Let $\sigma_i \in G$ map x_1 to x_i. It is an easy exercise for the reader to show that the right cosets $H\sigma_i$, $i = 1, 2, \ldots, n$, are distinct and fill out G. ■

Corollary A4.7. $|M_{23}| = 23\cdot 22\cdot 21\cdot 20\cdot 48$ *and* $|M_{22}| = 22\cdot 21\cdot 20\cdot 48$.

Note that $|M_{24}| = 24\cdot 23\cdot 22\cdot 21\cdot 20\cdot 48$ and recall the fact that M_{23} is a one-point stabilizer of M_{24} while M_{22} is a one-point stabilizer of M_{23}. ■

Proofs of the simplicity of M_{22}, M_{23} and M_{24} appear in [**15,** Prop. 3.6.3]. The first proofs of these facts were obtained by G. A. Miller in 1899 and 1900 [**37,** p. 169], [**90**], [**91**].

For further information on M_{24}, see [**25**], [**120**].

THE PROOF OF LEMMA 3.3

Recall that Lemma 3.3 stated that no symmetry in $\cdot 0$ has either prime order $p > 23$ or order $13 \cdot 23$. The first part was proved in the text. We now complete the proof.

Suppose λ is a symmetry of order $13 \cdot 23$ with matrix representation A. Let $P(x)$ be the minimum polynomial for A over the rational numbers.

It is known that the polynomial $x^n - 1$ is the product of the cyclotomic polynomials F_d, $d \mid n$, and that these are irreducible over the rationals (see [**66**, pp. 298, 299]). In particular,

$$\begin{aligned} x^{13 \cdot 23} - 1 &= F_1 \cdot F_{13} \cdot F_{23} \cdot F_{13 \cdot 23} \\ &= (x-1)(x^{12} + x^{11} + \cdots + 1) \\ &\quad \times (x^{22} + x^{21} + \cdots + 1) \\ &\quad \times (x^{12 \cdot 22} + \cdots + 1). \end{aligned}$$

Since A satisfies the equation $x^{13 \cdot 23} - 1 = 0$, and since the degree of $P(x)$ is at most the degree of the characteristic equation of A, the minimal polynomial for A can have only $x - 1$, $x^{12} + x^{11} + \cdots + 1$, or $x^{22} + x^{21} + \cdots + 1$ as factors. Thus $P(x)$ must be one of the following:

1. $x - 1$
2. $x^{12} + x^{11} + \cdots + 1$
3. $(x - 1)(x^{12} + x^{11} + \cdots + 1) = x^{13} - 1$

4. $x^{22} + x^{21} + \cdots + 1$
5. $(x - 1)(x^{22} + x^{21} + \cdots + 1) = x^{23} - 1.$

Note that $P(x)$ cannot have repeated factors since $x^{13\cdot 23} - 1$ does not.

Thus A satisfies one of the equations $x - 1 = 0$, $x^{13} - 1 = 0$, or $x^{23} - 1 = 0$, contradicting the assumption that it has order $13 \cdot 23$.

APPENDIX **6**

THE SPORADIC SIMPLE GROUPS

The main feature of this appendix is Table A6.1 which lists the twenty-six sporadic simple groups. Most of the information is taken directly from the corresponding tables in [**37**, pp. 172–176] and [**65**, p. 709].

As the table clearly shows, about the only items in complete agreement are the group orders. We have taken the dates of discovery from [**65**] as they indicate dates of initial discovery and not necessarily of the completed proof. The superscripts at the end of the discoverer(s') name(s) are used whenever there was a discrepancy. These correspond to the cited references.

Table A6.1 appears on pages 212–215.

TABLE A6.1.—

Group Notation	Discoverer(s)	Date
M_{11}	Mathieu[65] Mathieu-Cole[37]	1861
M_{12}	Mathieu[65] Mathieu-Miller[37]	1861
M_{22}	Mathieu[65] Mathieu-Miller[37]	1873
M_{23}	Mathieu[65] Mathieu-Miller[37]	1873
M_{24}	Mathieu[65] Mathieu-Miller[37]	1873
J or Ja or J_1	Janko	1965
$HaJW$ or J_2 or HaJ or J_1	Hall and Wales[65] Hall-Janko[37]	1967
HiS	D. Higman and Sims	1967
McL	Mclaughlin	1968
Sz or Suz	Suzuki	1968
$HJMcK$ or J_3 or HJM or J_2	G. Higman and McKay[65] Hall-Janko-McKay[37] Janko-Higman and McKay[20]	1968
$\cdot 1$ or Co_1	Conway[65] Conway-Thompson[37]	1968
$\cdot 2$ or Co_2	Conway[65] Conway-Thompson[37]	1968
$\cdot 3$ or Co_3	Conway[65] Conway-Thompson[37]	1968

—Continued

Order
$2^4 3^2 5 \cdot 11 = 7{,}920$ $2^6 3^3 5 \cdot 11 = 95{,}040$ $2^7 3^2 5 \cdot 7 \cdot 11 = 443{,}520$ $2^7 3^2 5 \cdot 7 \cdot 11 \cdot 23 = 10{,}200{,}960$ $2^{10} 3^3 5 \cdot 7 \cdot 11 \cdot 23 = 244{,}823{,}040$
$2^3 3 \cdot 5 \cdot 7 \cdot 11 \cdot 19 = 175{,}560$
$2^7 3^3 5^2 7 = 604{,}800$
$2^9 3^2 5^3 7 \cdot 11 = 44{,}352{,}000$
$2^7 3^6 5^3 7 \cdot 11 = 898{,}128{,}000$
$2^{13} 3^7 5^2 7 \cdot 11 \cdot 13 = 448{,}345{,}497{,}600$
$2^7 3^5 5 \cdot 17 \cdot 19 = 50{,}232{,}960$
$2^{21} 3^9 5^4 7^2 11 \cdot 13 \cdot 23 = 4{,}157{,}771{,}806{,}543{,}360{,}000$ $2^{18} 3^6 5^3 7 \cdot 11 \cdot 23 = 42{,}305{,}421{,}312{,}000$ $2^{10} 3^7 5^3 7 \cdot 11 \cdot 23 = 495{,}766{,}656{,}000$

TABLE A6.1.—

Group Notation	Discoverer(s)	Date
He or *HHMcK* or *HHM*	Held, G. Higman and McKay[65]	1968
$M(22)$ or Fi_{22} or F_{22}	Fischer	1969
$M(23)$ or Fi_{23} or F_{23}	Fischer	1969
$M(24)'$ or Fi'_{24} or Fi_{24} or F_{24}	Fischer	1969
Ly or *LyS*	Lyons-Sims	1970
R or *RCW* or *Rud*	Rudvalis-Conway-Wales[65] Rudvalis[37]	1972
O'N or *O'NS*	O'Nan-Sims	1973
F or *FLS* or *B* or F_2	Fischer and Leon-Sims[65] Fischer[37]	1973
T or F_3 or *E*	Thompson-Smith[65] Fischer-Smith-Thompson[37]	1974
HaCNS or F_5 or *F*	Harada-Conway-Norton-Smith[65] Fischer-Smith[37] Harada-Norton and Smith[20]	1974
M or F_1	Fischer Fischer-Greiss	1974
J_4	Janko Norton-Parker-Benson-Conway-Thackray	1975

—Continued

Order
$2^{10}3^3 5^2 7^3 \cdot 17 = 4{,}030{,}387{,}200$
$2^{17}3^9 5^2 7 \cdot 11 \cdot 13 = 64{,}561{,}751{,}654{,}400$ $2^{18}3^{13}5^2 \cdot 7 \cdot 11 \cdot 13 \cdot 17 \cdot 23 = 4{,}089{,}470{,}473{,}293{,}004{,}800$ $2^{21}3^{16}5^2 7^3 11 \cdot 13 \cdot 17 \cdot 23 \cdot 29 = 1{,}255{,}205{,}709{,}190{,}661{,}721{,}292{,}800$
$2^8 3^7 5^6 7 \cdot 11 \cdot 31 \cdot 37 \cdot 67 = 51{,}765{,}179{,}004{,}000{,}000$
$2^{14}3^3 5^3 7 \cdot 13 \cdot 29 = 145{,}926{,}144{,}000$
$2^9 3^4 5 \cdot 7^3 11 \cdot 19 \cdot 31 = 460{,}815{,}505{,}920$
$2^{41}3^{13}5^6 7^2 11 \cdot 13 \cdot 17 \cdot 19 \cdot 23 \cdot 31 \cdot 47 \approx 4.15 \times 10^{33}$
$2^{15}3^{10}5^3 7^2 13 \cdot 19 \cdot 31 = 90{,}745{,}943{,}887{,}872{,}000$
$2^{14}3^6 5^6 7 \cdot 11 \cdot 19 = 273{,}030{,}912{,}000{,}000$
$2^{46}3^{20}5^9 7^6 11^2 13^3 17 \cdot 19 \cdot 23 \cdot 29$ $\cdot 31 \cdot 41 \cdot 47 \cdot 59 \cdot 71 \approx 8.08 \times 10^{53}$
$2^{21}3^3 5 \cdot 7 \cdot 11^3 23 \cdot 29 \cdot 31 \cdot 37 \cdot 43 = 86{,}775{,}571{,}046{,}077{,}562{,}880$

BIBLIOGRAPHY

1. Abramson, N., *Information Theory and Coding*, McGraw-Hill, Inc., New York, 1963.

2. Alexanderson, G. L., George Pólya Interviewed on His Ninetieth Birthday, *Two-Year College Math. J.*, 10 (1979), 13–19.

3. Alt, F. L., A Bell Telephone Laboratories' Computing Machine—I, *Math. Tables and Other Aids to Computation* (*Math. Comput.*), 3 (1948–49), 1–13.

4. Alt, F. L., A Bell Telephone Laboratories' Computing Machine—II, *Math. Tables and Other Aids to Computation* (*Math. Comput.*), 3 (1948–49), 69–84.

5. Anderson, I., *A First Course in Combinatorial Mathematics*, Oxford University Press, London, 1974.

6. Andrews, E. G., Telephone Switching and the Early Bell Laboratories Computers, *Bell System Tech. J.*, 42:1 (1963), 341–353.

7. Assmus, E. F., Jr., and H. F. Mattson, Jr., On Tactical Configurations and Error-Correcting Codes, *J. Combinatorial Theory*, 2 (1967), 243–257.

8. Assmus, E. F., Jr., H. F. Mattson, Jr., and R. Turyn, *Cyclic Codes, Report AFCRL-66-348* (*AD-634-989*), Air Force Cambridge Research Labs., Bedford, Massachusetts, April 28, 1966.

9. Bannai, E., and N. J. A. Sloane, Uniqueness of Certain Spherical Codes, *Canad. J. Math.*, 33 (1981), 437–449.

10. Bartee, T. C., *Digital Computer Fundamentals*, third ed., McGraw-Hill Book Co., New York, 1972.

11. *Bell System Tech. J.*, 29 (1950), 294.

12. Berlekamp, E. R., *Algebraic Coding Theory*, McGraw-Hill Book Co., New York, 1968.

13. Berlekamp, E. R., *Key Papers in the Development of Coding Theory*, I.E.E.E. Press, New York, 1974.

14. Best, M. R., A. E. Brouwer, F. J. MacWilliams, A. M. Odlyzko, and N. J. A. Sloane, Bounds for Binary Codes of Length Less than 25, *I.E.E.E. Trans. Inform. Theory*, IT-24 (1978), 81–93.

15. Biggs, N., *Finite Groups of Automorphisms*, London Math. Soc. Lecture Note Series, No. 6, Cambridge Univ. Press, London, 1971.

16. Birkhoff, G., and S. Mac Lane, *A Survey of Modern Algebra*, 4th ed., Macmillan Pub. Co., Inc., New York, 1977.

17. Blake, I. F., *Algebraic Coding Theory: History and Development*, Dowden, Hutchinson and Ross, Stroudsburg, Pennsylvania, 1973.

18. Blichfeldt, H. F., The Minimum Values of Quadratic Forms and the Closest Packing of Spheres, *Math. Ann.*, 101 (1929), 605–608.

19. Blichfeldt, H. F., The Minimum Values of Positive Quadratic Forms in Six, Seven and Eight Variables, *Math. Z.*, 39 (1935), 1–15.

20. Brauer, R., Blocks of Characters and Structure of Finite Groups, *Bull. Amer. Math. Soc.* (*N.S.*), 1 (1979), 21–38.

21. Carmichael, R. D., *Introduction to the Theory of Groups of Finite Order*, Dover Pub., Inc., New York, 1956.

22. Cocke, J., Lossless Symbol Coding with Nonprimes, *I.R.E.* (*I.E.E.E.*) *Trans. Inform. Theory*, 5 (1959), 33–34.

23. Conway, J. H., A Perfect Group of Order 8,315,553,613,086,720,000 and the Sporadic Simple Groups, *Proc. Nat. Acad. Sci. U.S.A.*, 61 (1968), 398–400.

24. Conway, J. H., A Group of Order 8,315,553,613,086,720,000, *Bull. London Math. Soc.*, 1 (1969), 79–88.

25. Conway, J. H., Three Lectures on Exceptional Groups, Chapter VII of: M. B. Powell and G. Higman, *Finite Simple Groups*, Academic Press, New York, 1971.

26. Conway, J. H., Phone call, February 10, 1978.

27. Conway, J. H., Phone call, April 6, 1978.

28. Conway, J. H., Phone call, July 30, 1978.

29. Coxeter, H. S. M., *Regular Polytopes*, Pitman Pub. Corp., New York, 1948.

30. Coxeter, H. S. M., An Upper Bound for the Number of Equal Nonoverlapping Spheres That Can Touch Another of the Same Size, *Proc. Sympos. Pure Math.*, 7 (1963), 53–71, Amer. Math. Soc., Providence, Rhode Island.

31. Delsarte, P., and J.-M. Goethals, Unrestricted Codes with the Golay Parameters Are Unique, *Discrete Math.*, 12 (1975), 211–224.

32. Denniston, R. H. F., Some New 5-designs, *Bull. London Math. Soc.*, 8 (1976), 263–267.

33. Denniston, R. H. F., The Problem of the Higher Values of t, *Ann. Discrete Math.*, 7 (1980), 65–70.

34. Doyen, J., and A. Rosa, An Updated Bibliography and Survey of Steiner Systems, *Ann. Discrete Math.*, 7 (1980), 317–349.

35. Elias, Peter, Coding for Noisy Channels, *I.R.E.* (*I.E.E.E.*) *Conv. Rec.*, 3 (1955), 37–46.

35A. Falk, J. W., Letter to B. D. Holbrook, August 19, 1983.

36. Fisher, R. A., The Theory of Confounding in Factorial Experiments in Relation to the Theory of Groups, *Ann. Eugenics* (*Human Genetics*), 11 (1941-42), 341-353.

37. Gallian, J. A., The Search for Finite Simple Groups, *Math. Mag.*, 49 (1976), 163-179.

38. Gallian, J. A., Classification of Finite Simple Groups Completed, *MAA Focus*, 1 (1981), 3, 7.

39. Golay, M. J. E., Letter to C. E. Shannon, January 20, 1949.

40. Golay, M. J. E., Notes on Digital Coding, *Proc. I.R.E.* (*I.E.E.E.*), 37 (1949), 657.

41. Golay, M. J. E., Binary Coding, *Trans. I.R.E.* (*I.E.E.E.*), PGIT-4 (1954), 23-28.

42. Golay, M. J. E., Notes on the Penny-Weighing Problem, Lossless Symbol Coding with Nonprimes, etc., *I.R.E.* (*I.E.E.E.*) *Trans. Inform. Theory*, IT-4 (1958), 103-109.

43. Golay, M. J. E., Anent Codes, Priorities, Patents, etc., *Proc. I.E.E.E.*, 64 (1976), 572.

44. Golay, M. J. E., Letter, October 26, 1977.

45. Golay, M. J. E., Phone call, December 2, 1977.

46. Golay, M. J. E., Phone call, May 18, 1978.

47. Golomb, S. W., and E. C. Posner, Rook Domains, Latin Squares, Affine Planes, and Error-Distributing Codes, *I.E.E.E. Trans. Inform. Theory*, IT-10 (1964), 196-208.

48. Gorenstein, D., The classification of Finite Simple Groups, *Bull. Amer. Math. Soc.* (*N.S.*), 1 (1979), 43-199.

49. Hamming, R. W., *Self-Correcting Codes—Case 20878*, *Memorandum 1130-RWH-MFW*, Bell Telephone Laboratories, July 27, 1947.

50. Hamming, R. W., *A Theory of Self-checking and Self-correcting Codes—Case 20878*, *Memorandum 48-110-31*, Bell Telephone Laboratories, June 10, 1948.

51. Hamming, R. W., *Single Error-Correcting Codes—Case 20878*, *Memorandum 48-110-52*, Bell Telephone Laboratories, September 6, 1948.

52. Hamming, R. W., *Error Detecting and Error Correcting Codes*, Bell Telephone Laboratories, April 12, 1949.

53. Hamming, R. W., *Error Detecting and Error Correcting Codes*, Bell Telephone Laboratories, April 12, 1949—HV.

54. Hamming, R. W., *Error Detecting and Error Correcting Codes*, Bell System Tech. J., 29 (1950), 147-160.

55. Hamming, R. W., *Error Detecting and Correcting Codes*, Bell Lab. Record, 28 (1950), 193-198.

56. Hamming, R. W., Interview, February 3-4, 1977.

57. Hamming, R. W., Phone call, March 28, 1978.

58. Hamming, R. W., Letter, April 5, 1978.

59. Hamming, R. W., *Coding and Information Theory*, Prentice-Hall, Inc., Englewood Cliffs, New Jersey, 1980.

60. Hamming, R. W., and B. D. Holbrook, U.S. Patent No. 2,552,629, May 15, 1951.

61. Hartshorne, R., *Foundations of Projective Geometry*, Mathematics Lecture Note Series No. 10, Advanced Book Program, W. A. Benjamin, Inc., Reading, Massachusetts, 1967.

62. Herstein, I. N., *Topics in Algebra*, Blaisdell Pub. Co., Waltham, Massachusetts, 1964.

63. Holbrook, B. D., *Circuits for Self-Correcting Codes—Case 37896, Memorandum 3140-BDH-ML*, Bell Telephone Laboratories, February 1, 1949.

64. Holbrook, B. D., *Notes on Circuits for Self-Correcting Codes—Case 37896, Memorandum 3140-BDH-UH*, Bell Telephone Laboratories, August 9, 1949.

64A. Holbrook, B. D., Phone call, January 25, 1983.

65. Hurley, J. F., and A. Rudvalis, Finite Simple Groups, *Amer. Math. Monthly*, 84 (1977), 693–714.

66. Hungerford, T. W., *Algebra*, Holt, Rinehart and Winston, Inc., New York, 1974.

67. *I.R.E. (I.E.E.E.) Trans. Inform. Theory*, IT-4 (1958), 131.

68. Janko, Z., A New Finite Simple Group with Abelian 2-Sylow Subgroups, *Proc. Nat. Acad. Sci. U.S.A.*, 53 (1965), 657–658.

69. Janko, Z., A New Finite Simple Group with Abelian 2-Sylow Subgroups and Its Characterization, *J. Algebra*, 3 (1966), 147–186.

70. Julin, D., Two Improved Block Codes, *I.E.E.E. Trans. Inform. Theory*, 11 (1965), 459.

71. Klein, F., *Lectures on the Icosahedron and the Solution of Equations of the Fifth Degree*, Dover Pub., Inc., New York, 1956.

72. Leech, J., Some Sphere Packings in Higher Space, *Canad. J. Math.*, 16 (1964), 657–682.

73. Leech, J., Letter to the Editor, *Canad. J. Math.*, September 15, 1965.

74. Leech, J., Notes on Sphere Packings, *Canad. J. Math.*, 19 (1967). 251–267.

75. Leech, J., Letter, September 20, 1977.

76. Leech, J., Letter, October 28, 1977.

77. Leech, J., Letter, January 6, 1978.

78. Leech, J., Letter, April 10, 1978.

79. Leech, J., Letter, July 18, 1978.

80. Leech, J., Letter, August 21, 1978.

81. Leech, J., Letter, October 31, 1978.

82. Leech, J., Letter, June 3, 1980.

83. Leech, J., and N. J. A. Sloane, Sphere Packings and Error-Correcting Codes, *Canad. J. Math.*, 23 (1971), 718–745.

84. MacWilliams, F. J., An Historical Survey, in: H. B. Mann., ed., *Error-Correcting Codes*, John Wiley and Sons, Inc., New York, 1968.

85. MacWilliams, F. J. and N. J. A. Sloane, *The Theory of Error-Correcting Codes*, North-Holland Pub. Co., Amsterdam, 1977.

86. Mathieu, É., Mémoire sur L'Étude des Fonctions de Plusieurs Quantités, sur la Manière de les Former et sur les Substitutions Qui les Laissent Invariables, *J. Math. Pures Appl.*, *Ser. 2*, 6 (1861), 241–274.

87. Mathieu, É., Sur la Fonction Cinq Fois Transitive de 24 Quantités, *J. Math. Pures Appl.*, *Ser. 2*, 18 (1873), 25–46.

88. May, K. O., Historiographic Vices II. Priority Chasing, *Historia Math.*, 2 (1975), 315–317.

89. McKay, J., Phone call, February 14, 1978.

90. Miller, G. A., On the Simple Groups Which Can Be Represented as Substitution Groups That Contain Cyclical Substitutions of a Prime Degree, *Amer. Math. Monthly*, 6 (1899), 102–103.

91. Miller, G. A., Sur Plusieurs Groupes Simples, *Bull. Soc. Math. France*, 28 (1900), 266–267.

92. Mills, W. H., A New 5-design, *Ars Combinatoria*, 6 (1978), 193–195.

93. Muller, D. E., Application of Boolean Algebra to Switching Circuit Design and to Error Detection, *I.R.E. (I.E.E.E.) Trans. Electronic Comput.*, EC-3 (1954), 6–12.

94. *Notices Amer. Math. Soc.*, 26 (1979), 53.

95. Odlyzko, A. M., and N. J. A. Sloane, New Bounds on the Number of Unit Spheres That Can Touch a Unit Sphere in *n* Dimensions, *J. Combinatorial Theory*, *Series A*, 26 (1979), 210–214.

96. Paige, L. J., A Note on the Mathieu Groups, *Canad. J. Math.*, 9 (1957), 15–18.

97. Paige, L. J., Interview, October 27, 1977.

98. Paige, L. J., and E. H. Spanier, *Economical Substantialism Mod 2*, SCAMP Working Paper No. 20, August 24, 1955.

99. Paley, R. E. A. C., On Orthogonal Matrices, *J. Math. Physics* (*Studies in Appl. Math.*), 12 (1933), 311–320.

100. Peterson, W. W., *Error-Correcting Codes*, M.I.T. Press, Cambridge, Massachusetts, 1961.

101. Pless, V., On the Uniqueness of the Golay Codes, *J. Combinatorial Theory*, 5 (1968), 215–228.

102. Reed, I. S., A Class of Multiple-Error-Correcting Codes and the Decoding Scheme, *I.R.E. (I.E.E.E.) Trans.*, PGIT-4 (1954), 38–49.

103. Rispin, E., Notice of approval for publication of R. W. Hamming's Error Detecting and Error Correcting Codes, February 7, 1950.

104. Rogers, C. A., The Packing of Equal Spheres, *Proc. London Math. Soc., Ser. 3*, 8 (1958), 609–620.

105. Rogers, C. A., *Packing and Covering*, Cambridge Tracts in Mathematics and Mathematical Physics No. 54, Cambridge U. Press, London, 1964.

106. Rotman, J. J., *The Theory of Groups: an Introduction*, Allyn and Bacon, Inc., Boston, Massachusetts, 1965.

107. Schreier, O., Über die Erweiterung von Gruppen I, *Monatsh. Math. Phys.*, 34 (1926), 165–180.

108. Shannon, C. E., A Mathematical Theory of Communication, *Bell System Tech. J.*, 27 (1948), 379–423, 623–656.

109. Shannon, C. E., Telephone call, April 6, 1978.

110. Shannon, C. E., and W. Weaver, *The Mathematical Theory of Communication*, The Univ. of Ill. Press, Urbana, Illinois, 1949.

111. Sloane, N. J. A., Letter to J. Leech, October 23, 1969.

112. Sloane, N. J. A., Binary Codes, Lattices, and Sphere-Packings, pp. 117–164 of: P. J. Cameron, *Combinatorial Surveys*, Academic Press, London and New York, 1977.

113. Sloane, N. J. A., Self-dual Codes and Lattices, pp. 273–308 of: D. K. Ray-Chaudhuri, *Relations Between Combinatorics and Other Parts of Mathematics, Proc. Symp. Pure Math., xxxiv*, Amer. Math. Soc., Providence, Rhode Island, 1979.

114. Spanier, E. H., *Substantialism for Small Values*, SCAMP Working Paper No. 4, July 21, 1955.

115. Spanier, E. H., *Linear Substantialism*, SCAMP Working Paper No. 6, July 25, 1955.

116. Spanier, E. H., Interview, November 10, 1977.

117. Stein, S. K., Algebraic Tiling, *Amer. Math. Monthly*, 81 (1974), 445–462.

118. Sylvester, J. J., Thoughts on Inverse Orthogonal Matrices, Simultaneous Sign-successions, and Tessellated Pavements in Two or More Colours, with Applications to Newton's Rule, Ornamental Tile-Work, and the Theory of Numbers, *Philos. Mag., Ser. 4*, 34 (1867), 461–475.

119. Taussky, O., and J. Todd, Covering Theorems for Groups, *Ann. Soc. Polonaise de Math.*, 21 (1948), 303–305.

120. Todd, J. A., A Representation of the Mathieu Group M_{24} as a Collineation Group, *Ann. Mat. Pura Appl.*, 71 (1966), 199–238.

121. van der Waerden, B. L., *Algebra, Vols. 1 and 2*, The Frederick Ungar Pub. Co., Inc., New York, 1970.

122. van Lint, J. H., A Survey of Perfect Codes, *Rocky Mountain J. Math.*, 5 (1975), 199–224.

123. Watson, G. L., The Number of Minimum Points of a Positive Quadratic Form, *Dissertationes Math.* (*Rozprawy Mat.*), 84 (1971), 5–42.

124. Weaver, W., The Mathematics of Communication, *Sci. Amer.*, 181 (1949), 11–15.

125. Williams, S. B., Bell Telephone Laboratories' Relay Computing System, *Ann. Harvard Computation Lab.*, 16 (1947), 40–68.

126. Witt, E., Die 5-fach Transitiven Gruppen von Mathieu, *Abh. Math. Sem. Univ. Hamburg*, 12 (1938), 256–264.

127. Witt, E., Über Steinersche Systeme, *Abh. Math. Sem. Univ. Hamburg*, 12 (1938), 265–275.

128. Zaremba, S. K., 1952, Covering Problems Concerning Abelian Groups, *J. London Math. Soc.*, 27 (1952), 242–246.

INDEX

(**Note:** Underlining the page number indicates definition.)